T0181315

AutoUni – Schriftenreihe

Volume 159

Reihe herausgegeben von

Volkswagen Aktiengesellschaft, AutoUni, Volkswagen Aktiengesellschaft, Wolfsburg, Germany

Volkswagen bietet Wissenschaftlern und Promovierenden der Volkswagen AG die Möglichkeit, ihre Forschungsergebnisse in Form von Monographien und Dissertationen im Rahmen der „AutoUni Schriftenreihe" kostenfrei zu veröffentlichen. Die „AutoUni Schriftenreihe" ist eine Schriftenreihe der Volkswagen Group Academy, in der Dissertationen aus dem Volkswagen Doktorandenprogramm veröffentlicht werden. Über die Veröffentlichung in der Schriftenreihe werden die Resultate nicht nur für alle Konzernangehörigen, sondern auch für die Öffentlichkeit zugänglich.

Volkswagen offers scientists and PhD students of Volkswagen AG the opportunity to publish their scientific results as monographs or doctor's theses within the "AutoUni Schriftenreihe" free of cost. „AutoUni Schriftenreihe" is a series from Volkswagen Group Academy, in which dissertations from the Volkswagen doctoral program are published. The publication within the "AutoUni Schriftenreihe" makes the results accessible to all Volkswagen Group members as well as to the public.

Reihe herausgegeben von / Edited by
Volkswagen Aktiengesellschaft
Brieffach 1358
D-38436 Wolfsburg

„Die Ergebnisse, Meinungen und Schlüsse der im Rahmen der „AutoUni Schriftenreihe" veröffentlichten Dissertationen sind allein die der Doktorand*innen."

More information about this series at https://link.springer.com/bookseries/15136

Fabian Kai Dietrich Noering

Unsupervised Pattern Discovery in Automotive Time Series

Pattern-based Construction of Representative Driving Cycles

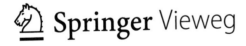

Fabian Kai Dietrich Noering
Braunschweig, Germany

Dissertation Technische Universität Braunschweig, 2021

ISSN 1867-3635   ISSN 2512-1154   (electronic)
AutoUni – Schriftenreihe
ISBN 978-3-658-36335-2   ISBN 978-3-658-36336-9   (eBook)
https://doi.org/10.1007/978-3-658-36336-9

Responsible Editor: Stefanie Eggert
This Springer Vieweg imprint is published by the registered company Springer Fachmedien
Wiesbaden GmbH part of Springer Nature.
The registered company address is: Abraham-Lincoln-Str. 46, 65189 Wiesbaden, Germany

Danksagung

Die vorliegende Doktorarbeit ist im Rahmen meiner Tätigkeit als Doktorand innerhalb der Group Innovation (ehemals Konzernforschung) der Volkswagen AG entstanden. Die dabei entstandenen Konferenzbeiträge und Veröffentlichungen ([35, 70–72]) fließen in diese Doktorarbeit mit ein.

Mein besonderer Dank gilt Herrn Prof. Dr. Frank Klawonn sowie Herrn Prof. Dr. Thomas M. Deserno für die wissenschaftliche Betreuung dieser Arbeit. In kurzen Intervallen führten wir immer wieder fachliche Diskussionen, die in der Ideenfindungsphase und darüber hinaus ungemein stimulierend wirkten. Die konstruktive Kritik, die sie mir und meiner Arbeit entgegenbrachten, hat mir bei der Bearbeitung sehr geholfen.

Seitens der Volkswagen AG möchte ich meinen internen Vorgesetzen für die Ermöglichung der Doktorarbeit sowie meinen Kollegen für die Unterstützung bei der Bearbeitung der Dissertation danken. Dabei möchte ich meinen Betreuer hervorheben, der mich mit seiner K1- typischen Art regelmäßig zur Innovation angetrieben hat. Ohne seine intiale visionäre Idee wäre diese Dissertion vermutlich nicht entstanden. Ein besonderer Dank gilt zudem all den Doktoranden, die mich während dieser herausfordernden Zeit begleitet haben und mir als Diskussions- und Sparringspartner dienten.

Weiterhin möchte ich meinen Studenten Yannik Schröder, Oliver Nick, Andreas Vieider und Andreas Braun danken, die ich im Rahmen meiner Doktorandenzeit betreuen durfte. Ihre Arbeit lieferte wertvolle Erkenntnisse, die sich auch in dieser Dissertation manifestierten.

Mein letzter Dank gilt meiner Familie und meinen Freunden, die mir sowohl in anstrengenden Phasen beistanden, als auch meine Erfolge mit mir teilten. Besonders danken möchte ich meinen Eltern Bärbel und Kai für ihre stetige Unterstützung und Johanna für ihre Geduld und die beizeiten notwendige Ablenkung.

Fabian Noering

Zusammenfassung

In dem letzten Jahrzehnt hat die unüberwachte Mustererkennung in Zeitreihen – die Identifikation von ähnlichen, wiederkehrenden, im Vorfeld nicht bekannten Teilsequenzen in langen, multivariaten Zeitreihen – immer mehr an Aufmerksamkeit in Forschung und Industrie gewonnen. Mustererkennung wurde in Bereichen wie der Seismologie, Medizin, Robotik oder der Musik bereits erfolgreich angewendet. Bis zu diesem Zeitpunkt wurde eine Anwendung auf automotive Zeitreihen nicht untersucht. Die vorliegende Dissertation schließt dieses Forschungsdesiderat, indem sie die besonderen Eigenschaften von Fahrzeug-Leerzeichen zu viel Sensor-Daten herausarbeitet und einen geeigneten Ansatz zur Mustererkennung vorschlägt. Zu diesen Eigenschaften gehören z. B. eine besonders hohe Dynamik und Diversität der Zeitreihen. Darüber hinaus bestehen Anforderungen bezüglich der Robustheit gegenüber Signalrauschen und der Fähigkeit mehr als 100 Mio. Datenpunkte in einer akzeptablen Rechenzeit verarbeiten zu können. Aufgrund der Diversität ist eine Methode notwendig, die Muster mit variabler Länge erkennen kann. Die vorgeschlagene Methode zur Mustererkennung erfüllt diese Anforderungen und stellt somit ein leistungsfähiges Werkzeug dar, um z. B. die Nutzung und Beanspruchung des Fahrzeugs oder einzelner Komponenten zu analysieren. Darüber hinaus kann die Methode als exploratives Data-Mining-Werkzeug für eine datengetriebene Unterstützung des Entwicklungsprozesses eingesetzt werden.

Um den Nutzen von Mustererkennung im automotiven Umfeld herauszustellen, wird die Methode zur Konstruktion von repräsentativen Fahrzyklen angewandt. Im Entwicklungsprozess werden solche Fahrzyklen häufig z. B. für vergleichbare Funktions- und Dauerlauftests von Komponenten genutzt. Die Idee der musterbasierten Zyklen-Konstruktion ist es, automatisiert die Fahrsituationen zu identifizieren, die notwendig sind, um den Großteil des Fahrzeugdatensatzes zu beschreiben. Darauf basierend werden zwei Ansätze zur Konstruktion und Identifikation von repräsentativen Fahrzyklen vorgestellt.

Abstract

In the last decade unsupervised pattern discovery in time series, i.e. the problem of finding recurrent similar subsequences in long multivariate time series without the need of querying subsequences, has earned more and more attention in research and industry. Pattern discovery was already successfully applied to various areas like seismology, medicine, robotics or music. Until now an application to automotive time series has not been investigated. This dissertation fills this desideratum by studying the special characteristics of vehicle sensor logs and proposing an appropriate approach for pattern discovery. These characteristics include high dynamics and diversity of the time series. Furthermore the method is required to be robust regarding signal noise and to process time series with far more than 100 mio. samples in an acceptable computing time. Because of the diversity there is a high need for a method that discovers patterns of variable length. Covering these requirements, the proposed pattern discovery is a powerful tool e.g. to analyze the usage and stress of single components or the vehicle as a whole. Furthermore it can be used as exploratory data mining tool for a data-driven support of the engineering process.

To prove the benefit of pattern discovery methods in automotive applications, the algorithm is applied to construct Representative Driving Cycles (RDC). The usage of RDCs is a common tool in the engineering-process e.g. for comparable functional and endurance testings of components. The underlying idea of pattern-based RDC construction is to automatically identify the real-life situations that are necessary to describe the majority of a large vehicle data set. Furthermore two approaches are proposed to either construct the RDC or identify the most representative trip.

Contents

Abbreviations

APF	Absolute Pattern Frequency
ATSP	Asymmetric Traveling Salesman Problem
CP	Characteristic Parameters
CoN	Concentration of Norm phenomenon
CO_2	Carbon Dioxid
DC	Driving Cycle
DM	Distance Metric
DTW	Dynamic Time Warping
ECG	Electrocardiogram
FDR	False Discovery Rate
FFT	Fast Fourier Transformation
FN	False Negative
FP	False Positive
GPU	Graphics Processing Unit
HBT	Hysteresis Breakthrough
HMM	Hidden Markov Models
LP	Linear Programming
MASS	Mueen's ultra-fast Algorithm for Similarity Search
MDL	Minimum Description Length
MK	Mueen-Keogh
MTF	Mean Tractive Force
NaN	Not a Number
NEDC	New European Driving Cycle
NN	Neural Network
NO_x	Nitrogen Oxide
PAA	Piecewise Aggregate Approximation

PAN	Pan Matrix Profile
PCA	Principle Component Analysis
PLA	Piecewise Linear Approximation
RDC	Representative Driving Cycle
RDE	Real Driving Emission
RDL	Reduced Description Length
RPC	Relative Pattern Coverage
SAX	Symbolic Aggregate Approximation
STAMP	Scalable Time series Anytime Matrix Profile
SVC	Symbolic Value Change
TICC	Toeplitz Inverse Covariance-based Clustering
TP	True Positive
TPR	True Positive Rate
TSAPSS	Time series Subsequence All-Pairs-Similarity-Search
TSP	Traveling Salesman Problem
UCR	University of California—Riverside
USD	Uniform Scaling Distance
VALMOD	Variable Length Motif Discovery
VSP	Vehicle Specific Power
WLTC	Worldwide harmonized Light vehicles Test Cycle

Symbols

The meaning of frequently used symbols or notations is given in the table below. If nothing else is stated in the context, the symbol in the first column has the meaning given in the second column.

Symbol	Description
T	Time series.
$T_{i,l}$	A subsequence, i.e. a subset from T of length l starting from index i.
P	A pattern P, i.e. a group of ≥ 2 similar subsequences.
$noSteps$	Number of bins/steps for the discretization of time series.
T_{disc}	A discretized time series.
$windowsize$	Size of the window for PAA. All samples within this window will be considered for the calculation of the average. In this context $windowsize$ equals the number of samples to average.
$stepsize$	In order to create a sliding window for the PAA, after averaging a window, the window is slid a certain number of steps (*stepsize*) to the right.
$denoiselevel$	Parameter for Wavelet denoising.
$hyst_{value}$	Relative value for the height of hysteresis boundary.
$h_{jumpinto}$	The height of the discretization interval that the symbolic value jumps into.
$hyst_{maxtime}$	Maximum time/length of hysteresis region.
hbt	Index of the hysteresis breakthrough.
svc	Index of the symbolic value change.

Symbol	Description
$logBasic$	Parameter for the logarithmic reduction technique that specifies the basis of the logarithm.
X_j^i	Results of the logarithmic reduction are symbols in the form of X_j^i with i being the magnitude of the number of represented symbols X and j the index of symbol X in the subsequence.
$MinCount$	Minimum number of occurrences of identical symbolic sequences in order to create a new pattern.
$MinPatternSize$	Parameter for the minimum symbolic length of a sequence, which is necessary for the sequence to be considered as a pattern.
$MaxPatternSize$	Parameter for the maximum symbolic length of a sequence, that can be discovered as pattern.
α	Threshold that defines a break condition for the greedy algorithm as a postprocessing technique for pattern discovery.
$overlap2Original$	Relative coverage of not yet represented parts of the time series for each pattern.
$overlap2Rperesented$	Relative coverage of already represented parts of the time series for each pattern.
$numMembers$	Absolute number of members of each pattern.
$symLength$	Symbolic length of a pattern.
$uniqueSymLength$	Number of different symbols within a pattern.
APF_j	Absolute pattern frequency of pattern j.
RPC_j	Relative pattern coverage of pattern j.
$meanLength_j$	Average length of pattern j.
$numPatterns$	Total number of discovered patterns.
$overlap$	$= \frac{|TP|}{|TP \cup FN \cup FP|} = \frac{|TP|}{|I_{pre} \cup I_{disc}|}$
TPR	$= \frac{|TP|}{|TP \cup FN|} = \frac{|TP|}{|I_{pre}|}$
FDR	$= \frac{|FP|}{|TP \cup FP|} = \frac{|FP|}{|I_{disc}|}$
PPV	$= \frac{|TP|}{|TP \cup FP|} = \frac{|TP|}{|I_{disc}|} = 1 - FDR$
$quality_{feat}$	A feature- and knowledge-based quality score for RDCs.
$quality_{hist}$	A quality score for RDCs based on two-dimensional histograms in velocity and longitudinal acceleration.
$quality_{binaryhist}$	A quality score for RDCs based on two-dimensional binary histograms in velocity and longitudinal acceleration especially for use case 1.

List of Figures

List of Tables

Introduction

1

1.1 Motivation

Unsupervised pattern discovery defines the task of finding similar recurring sub-sequences in large time series without having any knowledge about the shape, length or level of detail of these patterns. Existing pattern discovery algorithms have already been successfully applied to various research areas like seismology, medicine or music. In seismology they have been used to discover repeating earthquake sequences [84, 114]. Within the medical context the algorithms were applied to data logged from electrocardiogram (ECG) or electroencephalography (EEG) [7, 8, 16, 57]. For example in ECG data they were able to identify similar heart beats, which can be used for the detection and analysis of abnormal behavior. Beside that, pattern discovery algorithms were used to classify human motions captured by accelerometers of smartwatches [42] or by electromyography (EMG) [65]. The same algorithms can also be applied to audio data to identify similarities in different songs and subsequently fingerprinting them [87, 98].

Unfortunately until now, an application to automotive time series has only been investigated to a small extend in [34], where the authors identified turns in seven-dimensional driving data. In the scope of this work, automotive time series means data gained from vehicles. Even low-cost vehicles today are equipped with a large number of sensors, while this number is increasing with every new generation. In addition to sensor data, internally calculated information from the electronic control units (ECUs) can be logged as well. In total it is possible to record a few thousand signals per vehicle. Under consideration of these enormous amounts of potential data, a large variety of new applications is conceivable regarding the validation of hardware and software, as well as the analysis of stress, load and wearing of components. The latter analysis could be refined in order to do fault or aging detection.

F. K. D. Noering, *Unsupervised Pattern Discovery in Automotive Time Series*,
AutoUni – Schriftenreihe 159, https://doi.org/10.1007/978-3-658-36336-9_1

Beside that, the general usage of single vehicles or fleets can be analyzed. Hence, there is a need for highly automatized data mining methods like the unsupervised pattern discovery.

In comparison to data sets of other disciplines, automotive data is remarkably dynamic, diverse, high-dimensional and large (> 100 mio. samples). The diversity is characterized by the large amount of different situations that occur while using the vehicles in different regions, countries or climate zones. Beyond that, these situations are influenced by the driver itself. Due to the fact, that every person has a more or less unique driving style, the frequently occurring situations (and patterns) are varying. Even though automotive data sets are maybe not the most dynamic ones or have the largest sizes, the combination of these four properties makes the handling in pattern discovery applications challenging. At the same time one major requirement for the regular use of pattern discovery algorithms within the automotive development is an acceptable computing time despite the high complexity of the task.

This dissertation fills this gap by studying the special characteristics of automotive data and proposing an appropriate approach for pattern discovery. A central element of this approach is the reduction of complexity by compressing the time series. This compression is done by a discretization of signal values complemented by a reduction in the time domain. Subsequently, text mining algorithms can be applied to discover frequently occurring subsequences. In previous work [70, 72] this approach was already evaluated against other approaches and classified as suitable for fast pattern discovery in time series. In order to overcome weaknesses of discretization-based approaches this work includes the proposal of additional preprocessing methods for enhanced compression of the time series. Furthermore, a postprocessing technique is presented to automatically select relevant patterns based on a greedy strategy.

The approach enables the unsupervised discovery of patterns in large multidimensional time series including a selection mechanism to find a set of representative patterns. In general, the approach is applicable to any kind of time series, while all functions were developed to match the special requirements of automotive time series. In order to proof the effectiveness of pattern discovery, the method is applied to the construction of **representative driving cycles** (RDC).

Representative cycles are widely used to represent the real-life usage of technical devices in a compressed way. The need for representative cycles directly derives from the variety of the real-life usage and the large amounts of data that is available from the device under surveillance. From an engineering perspective the exact knowledge of how the devices are used in real life has big advantages regarding e.g. the design and control strategy of the device itself. Hence, methods to analyze the usage are of great interest. The construction of representative cycles is one data-driven method to transform the large amounts of data to manageable sizes and therefore reduce the effort of testing or add a data-driven perspective to the engineering process.

The definition of representativeness is a difficult task. In order to construct a representative cycle for a certain use case usually knowledge-based characteristic parameters are defined that capture the important features that need to be represented. Furthermore, a scoring function has to be defined that takes into account the characteristic parameters to assess the representativeness of cycles. For an RDC characteristic parameters could for example be statistical features (mean, max. , min. , standard deviation) of the velocity trajectory within a cycle or the energy consumption of the vehicle. Based on this representativeness function, different approaches can be applied to either choose or synthesize an RDC.

There are two major drawbacks of common approaches. First, the users have to put a lot of effort into defining representativeness, if they are confronted with a new use case. Secondly, common characteristic parameters only capture a limited reality due to the fact that they are highly simplified. Hence, it is questionable if these parameters are able to capture the real-life usage and at the same time are able to express the representativeness of a cycle.

This work proposes a new technique to derive representative cycles without the need of defining knowledge-based characteristic parameters for representativeness. In general, the idea is to discover the smallest set of patterns that in total covers the majority of the data set. Hence, the pattern set is representative for the data set. A representative cycle should contain all patterns not only once but as frequent as in the initial data set related to its total length. The representativeness is now defined by discovered patterns and their frequency. Hence, the effort of defining knowledge-based characteristic parameters can be reduced. As a consequence, this technique can be quickly adapted to fit the requirements of other representative cycle use cases. Using this pattern-based representativeness, this work introduces two approaches for the construction of representative cycles.

1.2   Structure of the Thesis

This thesis starts in Chapter 2 with an extensive literature review, beginning in Section 2.1 with a description of state of the art pattern discovery techniques taking into consideration the previously named challenges. Subsequently, in Section 2.2 common techniques for the construction of representative driving cycles are reviewed. Furthermore, in Section 2.3 the idea of pattern-based RDCs is presented. Chapter 3 introduces the reader to the unsupervised pattern discovery framework as the major contribution of the thesis. The chapter begins with a discussion of special requirements in automotive time series and the suitability of different pattern discovery approaches under consideration of these requirements in Section 3.1. Afterwards, Sections 3.2 to 3.5 describe the pattern discovery framework in detail, including functions for the reduction of complexity, the symbolic pattern discovery and the selection of relevant patterns. After laying the foundation of pattern discovery, Chapter 4 proposes the pattern-based RDC construction. In Section 4.1 the chapter begins with an introduction to pattern-based statistics, that is necessary for the following construction of RDCs. Two alternative approaches for the pattern-based construction of RDCs are proposed in Sections 4.2 and 4.3. The proposed approaches for pattern discovery and the construction of RDCs are evaluated in Chapter 5. First, Section 5.1 works out the effectiveness of the whole pattern discovery framework as well as single functions. Secondly, Section 5.2 shows the benefit of pattern-based RDC construction techniques in contrast to a common approach. Last but not least, the results are critically discussed within Chapter 6. Furthermore, the chapter gives an outlook to future research.

Related Work

2

This chapter includes an extensive review of related work in literature. As this work can be separated in two parts, i.e. the general method of unsupervised pattern discovery in time series and its application for representative driving cycles, this chapter will address both topics separately, too. Section 2.1 begins with a discussion of the term pattern, its range and diversity. This is necessary in order to align the expectations and the focus of this work. Beside this, the section gives an introduction to the topic of pattern discovery in time series and gives an overview of existing approaches. Additionally, various challenges and requirements are discussed. Section 2.2 gives an introduction into the field of representative driving cycles including common applications, the definition of representativeness and existing approaches for the construction. Finally, Section 2.3 points out the obvious, but yet not discovered, linkage of both topics.

2.1   Pattern Discovery in Time Series

Unsupervised pattern discovery in time series is informally defined as the identification of recurring similar subsequences within a time series, while the length of a subsequence is usually much shorter than the time series itself. As this method is unsupervised, an input regarding e.g. the shape of the pattern is not needed to discover important patterns within the time series. Depending on the background of the user the intuition of a pattern can vary a lot and therefore lead to misunderstandings. First of all the intuition of a pattern is influenced by the scope of the users problem, which can exemplarily be formulated as:

© The Author(s), under exclusive license to Springer Fachmedien
Wiesbaden GmbH, part of Springer Nature 2022
F. K. D. Noering, *Unsupervised Pattern Discovery in Automotive Time Series*,
AutoUni – Schriftenreihe 159, https://doi.org/10.1007/978-3-658-36336-9_2

- Discover the most frequent patterns in the data set.
- Discover rare patterns that are unusual or anomalous within the data set.
- Discover a set of patterns that covers the whole data set.

Beside the scope, the intuition of a pattern is influenced by the users requirements for similarity. In Table 2.1 different variances of similar subsequences are shown. While on the left side the original subsequence is shown, the right side shows a variation of the original subsequence, which might be considered as similar depending on the users criteria. Just to address a few of them:

- Are two subsequences similar to each other if they are just similar in shape? Are they required to be in the same value range and have the same amplitude? See Table 2.1 *b) offset* and *c) amplitudinal scaling*.
- Is one of the subsequences allowed to be stretched or compressed in time? Or is it necessary for them to be of equal length? See Table 2.1 *d) longitudinal scaling*.
- Are the members of a pattern required to be similar in multiple dimensions? Are the dimensions known in advance? Is it possible that the corresponding dimensions of a pattern are shifted? See Table 2.1 *g) multidimensional shifting*.

In general, the biggest challenge in pattern discovery is to find an approach with a similarity measure matching the intuition of the user, while at the same time enabling the method to produce results in acceptable computing time. Because of the diversity of requirements there is also a variety of approaches that solve the pattern discovery problem. However, two competitive approaches emerged to be the ones applied mostly, namely approaches based on *Matrix Profile* and approaches based on *discretization* of the time series. The strategy of Matrix Profile in general is the similarity calculation of every possible subsequence with every other subsequence. Because the brute force approach is obviously not suitable, due to high computational complexity, a variety of techniques has been proposed to reduce the effort. The approaches based on discretization usually try to reduce the complexity of the time series in advance by transforming real-valued time series into the discrete space. This has the advantage of speeding up the calculations, because the task of similarity calculation in most cases reduces to checking for identity of subsequences. Beyond that, there are many other approaches, that have not (yet) been applied as much as Matrix Profile or discretization-based approaches.

Table 2.1 Different possibilities of variances in the shape of patterns

Original	Varied subsequence	Variance
a		noise
b		offset
c		amplitudinal scaling
d		longitudinal scaling
e		linear drift
f		irregularities
g		multidimensional shifing

At first, this section treats the terminology of the phrase *pattern discovery* and its diversity in Section 2.1.1. Furthermore the scope of pattern discovery handled within this work is underlined. Considering this, in Section 2.1.2 and Section 2.1.3 the Matrix Profile and discretization-based approaches are described in more detail. Furthermore, in Section 2.1.4 other approaches are briefly pictured.

2.1.1   Terminology

As mentioned before, the term pattern discovery can be interpreted in many different ways. To create the necessary transparency, at first the fundamental terms *time series* as well as *subsequence* need to be defined:

Definition 1. *A time series* $T \in \mathbb{R}^{n \times d+1}$ *is a sequence of real-valued tuples* $t_i \in \mathbb{R}^{d+1} : T = [t_1, t_2, \ldots, t_n]$ *where d corresponds to the number of dimensions and n is the length of T. Every t_i includes signal values of every dimension* $[y_{1,i}, y_{2,i}, \ldots, y_{d,i}]$ *and a time stamp x_i, while all the time stamps from i to n are strongly monotonically increasing and the intervall between all x_i and x_{i+1} is usually constant.*

Definition 2. *A subsequence* $T_{i,l} \in \mathbb{R}^{n \times d+1}$ *of a time series T is a subset of values from T of length l starting from index i. This subsequences $T_{i,l}$ contains consecutive tuples* $[t_i, t_{i+1}, \ldots, t_{i+l-1}]$.

Following these definitions every time series T or subsequence $T_{i,l}$ can be *univariate* if $d = 1$ or *multivariate* if $d > 1$.

Pattern vs. Motif
One of the most used names in the wide field of pattern discovery in time series is *motif discovery*. Unfortunately, definitions of time series motifs, or just motifs, are not consistent in the literature. For example within the Matrix Profile approach it is defined as 'the most similar subsequence pair of a time series' [103], while Lin et al. described it as 'a pattern that consists of two or more similar subsequences based on some distance threshold' [52]. To solve this inconsistency this work assumes a definition of motifs based on Matrix Profile [103]:

Definition 3. *A time series motif is the most similar subsequence pair of a time series. Formally, $T_{a,l}$ and $T_{b,l'}$ is the motif pair iff $dist(T_{a,l}, T_{b,l'}) \leq dist(T_{i,l''}, T_{j,l'''}) \forall i, j, l'', l'''$ excluding all trivial matches and dist is a function that computes the distance between two subsequences.*

A *trivial match* is a subsequence that is matched to itself or to a subsequence largely overlapping with itself. Hence, a subsequence and its adjacent subsequence can not be considered as the motif pair. Note that the original definition of a motif is slightly adapted here, because of the limitation to equal length motifs. In general, two subsequences of a motif can be of unequal length if the corresponding similarity measure is able to cope with it. An algorithm that solves the problem of motif discovery can always be extended to find the k-best motifs like e.g. in [74]. The result is a set of k motifs with the highest similarity or smallest distance scores. In contrast to the term motif, a pattern is now defined by following the explanations of [52]:

Definition 4. *A time series pattern P is a group of m (with m $\geq$ 2) similar subsequences in a time series T excluding all trivial matches. A subsequence $T_{i,l}$, that is included in a pattern P, is called a member M. Formally, $P = [M_1, M_2, \ldots, M_m]$ with each M_x being a subsequence with a starting index i and a length l. This includes a distance threshold τ with $dist(M_x, M_y) \leq \tau \; \forall x, y \in [1, \ldots, m]$.*

Hence, the definition of a pattern includes the definition of a motif. Consequently an algorithm that solves the problem of motif discovery can most likely be extended to do pattern discovery.

Differentiation from other Terms

Beside the motif discovery there are more terms that are closely related to the discovery of patterns. First of all the authors of [104] described time series *discords* as the inverse of a motif, i.e. the subsequence pair with the highest distance. Another time series primitive is introduced in [100], namely time series *shapelets*. Shapelets are patterns that are discovered in multiple data sets, with these shapelets being representative for the class of the data set, while the class of a data set is known beforehand. Hence, if a certain shapelet can be identified in a data set, this data set can be assigned to a certain class. In [112] the authors proposed the term time series *chain*, which is a pattern that evolves over time. While two neighbored subsequences in a chain have a high similarity, the first and the last subsequence can differ a lot. The discovery of chains can for example be applied to analyze a drifting behavior in a system. The task of finding time series *snippets* was formalized in [43], which is inspired by request "show me a typical sequence of data". Snippets are usually much longer than patterns and have the additional property of representing a high proportion of the input time series. Two similar snippets can for example contain approximately the same patterns, but in a different order. Last but not least, the process of *segmentation* is also related to pattern discovery, because a segment can obviously be a pattern. Unfortunately, the term segmentation is overloaded in the literature. As described in [27], segmentation on the one hand refers to approximating signals by e.g. Piecewise Linear Approximation (PLA). On the other hand, segmenting methods are considered to be used for change point detection, meaning the long-term change of statistical properties in a time series. Beside this, supervised segmentation methods are mostly used to label time series data. Because of this diversity, this work uses the term segmentation more like a generic term for methods that split or label data sets.

Dimensionality

As already stated, the requirements for pattern discovery in automotive time series include the handling of multivariate time series. In comparison to the univariate case, the complexity of pattern discovery in multivariate time series can increase extremely fast. Suppose a pattern is defined in just two dimensions, there are 13 possibilities for the chronology of subdimensional subsequences as already described by Alan's interval algebra [4] (see Table 2.2). Furthermore, there is another challenge regarding subdimensional pattern discovery, which was addressed by the authors of [103]. They named three different queries: guided, constrained and unconstrained pattern search in multivariate time series. *Guided* search refers to the problem of finding k-dimensional patterns in a d-dimensional time series without a specification which k dimensions to use. The *constrained* search additionally treats a user input, which dimensions specifically to include or exclude. Beyond this, the *unconstrained* search describes the task of finding k-dimensional patterns in d-dimensional time series, but without the knowledge of k.

Because of the high complexity of multivariate pattern discovery most approaches reduce their work to simultaneously occurring subdimensional patterns. Solutions for guided, constrained and unconstrained search can be found in the literature, while the most publications handle the easiest type of constrained search with $k = d$. Hence, the query is finding multivariate patterns that happen simultaneously in all

Table 2.2 Alan's intervall algebra describing the temporal relation of two sequences (illustration based on [4])

Relation	Symbol	Symbol for Inverse	Pictorial Example
X before Y	<	>	XXX YYY
X equal Y	=	=	XXX YYY
X meets Y	m	mi	XXXYYY
X overlaps Y	o	oi	XXX YYY
X during Y	d	di	XXX YYYYYY
X starts Y	s	si	XXX YYYYYY
X finishes Y	f	fi	XXX YYYYYY

given dimensions. In the following, this simple type is named as *all-dimensional* pattern discovery.

2.1.2 Matrix Profile

The idea of *Matrix Profile* was introduced at the University of California—Riverside (UCR) by a collaboration of many researchers, first and foremost Eamonn Keogh, Abdullah Mueen, Chin-Chia Michael Yeh and Yan Zhu. In the literature it is therefore referred to as *UCR Matrix Profile*, or in this work simply as Matrix Profile. In general it was developed to identify exact motifs and discords in time series. Since 2016, when the first Matrix Profile paper was published, it grew to an open source framework for many time series analysis tasks and earned special attention in research and industry. Until today the authors wrote more than 20 publications regarding the optimization, extension and application of Matrix Profile, which are consecutively named as Matrix Profile I to XX. A general guide of what tasks the Matrix Profile can be applied to and how to use it can be found in [115]. Even though the task of motif discovery is, strictly speaking, not equal to pattern discovery, as described in Section 2.1.1, it is beneficial to cover this approach at this point. Especially because it can be extended to a pattern discovery method, which was already outlined and tested in prior work [70]. In this section at first the basic idea and history of Matrix Profile is described, which is followed by brief explanations of relevant extensions and optimizations.

Theory of Matrix Profile
The basic Matrix Profile, which was introduced in [104], is the result of a *Time series Subsequence All-Pairs-Similarity-Search* (TSAPSS), more precisely the similarity or distance calculation of every subsequence of length m with every other subsequence of the same length within one and the same time series T of length n ($n >> m$). To perform the TSAPSS, at first a distance profile is calculated for every possible subsequence query Q with the length m. An exemplary distance profile D is shown in Fig. 2.1. It describes the distance of the query Q and every other subsequence in T. Because Q itself is included in T, the distance profile value at the starting position i of Q must be zero, while the values before and after i will usually be close to zero. The latter statement is false only if the time series is just random noise. This property of the distance profiles needs to be corrected to avoid trivial matches. Therefore, the gray area in Fig. 2.1 is ignored. The Matrix Profile stores the minimum distance of every query to every other subsequence of length m as shown in Fig. 2.2. With every calculated distance profile the Matrix Profile can be

updated. Hence, the Matrix Profile contains the distance of every subsequence in T to its nearest neighbor. Additionally, a Matrix Profile index vector is needed to store the information about the location of the nearest neighbor. The maximum of the Matrix Profile now matches with the discord of T, while the minimum corresponds to the best motif.

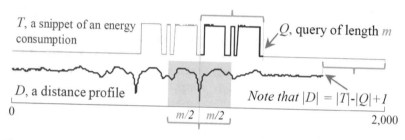

Figure 2.1 Distance profile D (bottom) of an exemplary time series T (top) and a query subsequence Q [104]

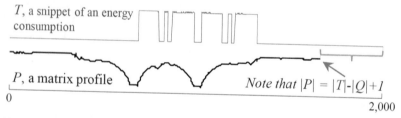

Figure 2.2 Matrix profile P (bottom) of an exemplary time series T (top) [104]

History of Matrix Profile

The problem of TSAPSS was already addressed in 2009, before the invention of the actual Matrix Profile, by Mueen and Keogh [67]. They described a first try to speed up the brute force calculation of the TSAPSS by utilizing the triangle inequality to reduce the amount of distance-calculations. Hence, they introduced the first exact non-brute-force fixed length motif discovery algorithm named MK (Mueen-Keogh). The authors of [73, 74] extended this idea to do variable length motif discovery by the use of MK for every possible motif length and an additional algorithm to group motifs together, select representatives and to score them. This is known as k-Best Motif Discovery algorithm (kBMD). In [105] they additionally proposed a speedup regarding previous publications including the use of the compression scheme Minimum Description Length (MDL) for the grouping of variable length motifs. Beyond that in [66] Mueen published another technique to speed up the brute force TSAPSS to discover maximally covering motifs of variable length. Based on a given minimal and maximal motif length the algorithm first calculates motifs of the minimal length by a smart brute force approach and then iteratively extends these for longer motifs.

Basic UCR Matrix Profile
In general, the UCR Matrix Profile focuses on the z-normalized Euclidean distance as a similarity measure. It preserves the shape of the subsequences, while being invariant to the absolute value of the subsequence by z-normalizing every single subsequence separately. Beyond that, the UCR Matrix Profile claims to be parameter free, but in most cases at least the subsequence length m has to be set by the user. Furthermore the authors strategically decided to fix some parameters for ease of usage, e.g. the area to avoid trivial matches is set to $m/2$ in [104]. In general the calculation of a Matrix Profile is computational expensive. To overcome the issue of brute-forcing through the TSAPSS problem to calculate a Matrix Profile, Mueen's ultra-fast Algorithm for Similarity Search (MASS) was introduced in [104]. It utilizes the Fast Fourier Transformation (FFT) to calculate a whole z-normalized Euclidean distance profile at once. Because of the highly optimized FFT algorithm, the MASS algorithm extremely reduces the computing time. Beside this, in previous work [70] an optimized method for standard (not normalized) Euclidean distance profile calculation was already proposed, which is even faster when it comes to calculating the exact Matrix Profile. One major advantage of the UCR Matrix Profile is fast convergence of the approximate Matrix Profile. Because the distance profiles are calculated in a random order instead of consecutively from the beginning of the time series, the authors achieved a faster convergence against the exact solution. This opens up the opportunity to terminate the computation at an early stage with a good approximate solution. If the exact solution is not needed, this approach saves a lot of computing time. Last but not least, the Matrix Profile can be computed in a highly distributed way due to the independence of the distance profiles, which is why the method is named *Scalable Time series Anytime Matrix Profile* (STAMP).

Extension to Multivariate Pattern Discovery
The authors of UCR Matrix Profile proposed an algorithm for multivariate motif discovery called mSTAMP in [103], which includes solutions for guided, constrained and unconstrained search. When calculating a multivariate distance profile of a query the algorithm simply calculates the distance profile for every dimension separately, which can afterwords be visualized as a $d \times (n - m + 1)$ matrix of distance profiles.

Every column is now being sorted, while keeping the information about the allocation of dimensions. For guided search of k-dimensional motifs in a d-dimensional
space the first k rows with the lowest distances are used for the Matrix Profile.
An additional vector, equivalent to the Matrix Profile index vector, containing the
information about the allocation of dimensions is created. For constrained search
the multivariate distance profile can also be adapted to exclude or include some
dimensions according to the users demands. For unconstrained search the problem reduces to finding the best k and then do guided search. Therefore the authors
proposed to run the mSTAMP algorithm separately with every possible k and then
evaluating the resulting Matrix Profiles. The idea is to find the k where the overall
Matrix Profile score significantly increases, which they solved by applying MDL.

As mentioned earlier, the motif discovery based on Matrix Profile can be extended
to discover patterns. Beside the solution proposed in [70], the UCR Matrix Profile
also offers a solution for that problem. In [101] they applied MDL to pick the best
motif candidates and the matching subsequences, which then can be called pattern
discovery. In order to do that they first apply STAMP to calculate the Matrix Profile.
Then their algorithm picks iteratively candidates, which correspond to the lowest
values in the Matrix Profile. These candidates are then tested to be a pattern or to
belong to an already existing pattern. This test is done by evaluating the bit costs that
can be saved when introducing it to be a new pattern or assigning it to an existing
pattern. These bit costs are calculated by the Reduced Description Length (RDL)
compression scheme, that evaluates the mismatch between two subsequences that
were discretized given a number of bits. The overall goal of the algorithm is the
minimization of the total bit costs, which is the memory that is needed to store
the patterns, their members and all the subsequences that were not assigned to a
pattern in consideration of the RDL compression. The algorithm stops if the total
bit costs can not be reduced by another candidate. The authors also proposed the
visualization of these patterns by Multi-Dimensional Scaling (MDS).

The XIXth part of the UCR Matrix Profile [61] proposes a similar technique, also
based on the MDL principle, but embedded in an interactive labeling system called
Like-Behaviors Labeling Routine (LBLR). It enables the user to edit automatically
discovered patterns by adding or removing subsequences.

Extension to Variable Length Motif Discovery
Up until now the UCR Matrix Profile only solves problems with fixed subsequence
length m, which is the only real parameter. Because of the high demand to be
truly parameter-free they proposed multiple solutions to this problem in the Matrix
Profile parts IV, X, XVI and XX. In Matrix Profile IV [102] they first proposed a
naive solution in the context of a streaming application for successive learning of

a motif-dictionary with variable length to predict labels in real-valued time series. They simply calculated multiple Matrix Profiles for a set of subsequence lengths and then tested the best motif candidates for their ability to predict the label. In Matrix Profile XVI [42] they indirectly solved the variable length problem by introducing a new segmentation type called semantic motif. It is based on the theory that a motif, like a handshake generated by an accelerometer of a smart-watch, can have a varying number of hand-pumps and still is a semantically important subsequence. Unfortunately the common approaches to find motifs are not able to find all those handshakes, which is why the authors introduced the semantic motif that consists of a prefix, a "don't care" region and a suffix. Within a "don't care" region small submotifs like a hand-pump are ignored and can therefore include multiple replicates. Because the length of this region is variable, the resulting motifs are of variable length, too. Nevertheless this is a highly specialized form of variable length motif discovery and is certainly not made for the general variable length problem.

A more general solution was proposed in Matrix Profile X [58] named Variable Length Motif Discovery (VALMOD). First of all they needed to introduce a distance measure that is normalized regarding the length of the subsequence, to be able to compare distances of different subsequence lengths. In general VALMOD still calculates the Matrix Profile for a set of different lengths $[l_{min}, l_{max}]$, that are aggregated to a Variable Length Matrix Profile (VALMP). The VALMP is still a vector like the standard Matrix Profile, i.e. containing at every index the minimum length-normalized distance to its nearest neighbor. But the corresponding subsequences can now have a length between l_{min} and l_{max}, while the subsequence and its nearest neighbor still have the same length. Therefore another meta vector is necessary, which stores the corresponding subsequence length equivalent to the Matrix Profile index. VALMOD optimizes the calculation of a VALMP by calculating lower bounding distance profiles for greater length subsequences based on smaller lengths.

A competing approach is proposed in Matrix Profile XX [60]. In contrast to Matrix Profile X it does not aggregate the Matrix Profile of different lengths to one vector, but to a matrix containing the Matrix Profiles of every evaluated length, which is called Pan Matrix Profile (PAN). Additionally it comes with a visualization tool for the user to quickly spot regions of variable length motifs. Beyond this they proposed a solution to reduce the previously required input of a minimal and maximal subsequence length in order to make the analysis truly parameter-free. While l_{min} is set to 2 by default, it is more difficult to identify a reasonable l_{max}. In general the first row in the PAN that is free from meaningful motifs is the upper bound. But in order to determine if there is something meaningful, a threshold needs to be fixed for the distance. Even though the algorithm is not truly parameter-free, the authors reason that this threshold, once been set to a reasonable value, is robust at least in one domain. In order to keep the computation time low the algorithm called Scalable Kinetoscopic Matrix Profile (SKIMP) uses the redundancy of Matrix Profiles of similar lengths by ordering the calculations for a faster convergence.

Optimization

As mentioned earlier one focus of the UCR Matrix Profile is the reduction of computing time to enable more complex analyses for various applications. This includes e.g. the ability to process larger time series, the scalability of calculations or the applicability to use cases with streaming requirements while maintaining the real-time computability. Therefore, Matrix Profile II, XI and XIV additionally introduce optimizations for the Matrix Profile tasks. In Matrix Profile II [114] the authors proposed a solutions to run the distance profile calculations on Graphics Processing Units (GPUs). In Matrix Profile XIV [116] they went another step further by bringing the calculations to commercial GPU clusters, which enables them to process time series with more than a quintillion (10^{18}) data samples. Beside this, Matrix Profile XI [113] proposes to run a function that pre-calculates an approximate Matrix Profile with a subsequent exact calculation that takes advantage of the approximate solution. The pre-function is based on the so called consecutive neighborhood preserving property, which says that in the Matrix Profile there will be consecutive indices that refer to consecutive subsequences as their nearest neighbors. Hence, this optimization is made for a faster convergence to the exact Matrix Profile.

Beyond Motifs and Patterns

Beyond the definition of motifs and patterns UCR Matrix Profile also offers solutions to other similar problems in Matrix Profile VII, VIII, XII, XIII, XVI and XV. Even though these extensions are not 100% relevant to the topic of pattern discovery itself, they can be highly interesting, because they solve problems that are encountered usually after applying pattern discovery. In Matrix Profile VII [112] the authors proposed a solution to discover time series chains, which are motifs that evolve over time. Semantic segmentation is a method introduced in Matrix Profile VIII [27] that is able to automatically mark a higher level behavior change based on motifs that were detected in the segments. Beside this in Matrix Profile XII [28] the authors invented MPdist, a distance measure based on the Matrix Profile to calculate distances of long variable length time series. MPdist is used in Matrix Profile XIII [43] to discover time series snippets, that are able to represent a high percentage of the input time series. In Matrix Profile XV [45] they offer a solution to find motifs in different data sets, which they call consensus motifs. Last but not least, a semantic motif finder was introduced in Matrix Profile XVI [42].

2.1.3   Discretization-based Approaches

The research in discretization-based pattern discovery has earned more and more attention since the first proposal of the Symbolic Aggregate Approximation algorithm (SAX) in [54] by Keogh and Lin. It combines the filtering of time series by a Piecewise Aggregate Approximation (PAA) with a sliding window and subsequently equi-probable discretization. By applying SAX or other discretization algorithms the complexity of the real-valued time series is reduced. Based on this symbolic time series many authors developed methods to find recurring patterns, initially with fixed length and later with variable length. This section begins with a review of algorithms for pattern discovery based on discrete time series, followed by techniques to enhance the algorithms like additional pre- and postprocessing methods, multiresolution discretization and the handling of multivariate time series.

Discovering Repeated Words
In 2003 Keogh et al. utilized the *random projection* algorithm to discover approximate fixed length patterns in discretized time series [12]. After discretizing the time series the random projection algorithm creates a matrix S of the size $(n-m+1 \times m)$ by a sliding window with each row corresponding to a word of length m in the discretized time series of length n. A word is a series of consecutive symbols in the discretized time series. By randomly masking columns of the matrix S and hashing the leftover symbols of each row, a collision matrix for each word in the time series can be calculated. Every time two masked rows in S are equal the corresponding cell in the collision matrix is increased by one. This process of masking, hashing and counting is repeated several times. Hence, the high values in the collision matrix refer to possible patterns. Keogh proposed to iteratively run through the cells with the highest collision counts, extract the motifs and double check the corresponding subsequences regarding their similarity based on the original time series. The same approach of random projection was also used in [95] and [63] and extended to perform multidimensional pattern discovery. In general this approach has the desired property that members of a pattern do not necessarily have identical symbolic representations. On the downside it is limited to fixed length analysis and has a high memory demand because of the collision matrix, which has the approximate size of n^2.

Tanaka and Uehara introduced algorithms based on the concept of MDL. While in [91] they described the discovery of patterns with fixed length, in [90] they extended this idea to variable length patterns. The pattern discovery algorithm based on MDL acts like a compression algorithm. Hence, its goal is to find patterns that best compress the time series. A measure for the compression is the description length.

This measure is minimized by selecting patterns that are occurring as frequently and are as complex as possible. The MDL is calculated for every possible variable length pattern under the constraint of identical symbolic sequence and a maximum Euclidean distance, which is afterwords used to extract the best patterns.

The Sequitur algorithm, first introduced in [69] to extract grammatical rules from texts, follows a similar principle like the MDL, namely compressing the symbolic time series. It extracts a hierarchical structure from the symbolic time series and at the same time compresses it by replacing every discovered pattern by a new symbol. Hence, its memory usage reduces with every pattern while running the algorithm. In more detail, the algorithm iteratively searches for recurring symbolic sequences of length 2, named digrams. If a match is found, the digrams are replaced by a new symbol. At the same time the digram is saved as a rule in a dictionary. Because a rule can also contain a lower level rule, the algorithm is able to discover patterns with variable length. The combination of SAX and Sequitur is one of the most applied time series pattern discovery algorithms since its first proposal in [57]. They applied it and then ranked each pattern by their frequency, rule length and variation to be able to extract only the interesting patterns. In [52] Lin et al. used it and added another preprocessing method, that enables the Sequitur to discover time scaled patterns. Furthermore a visualization tool was developed and published in [81, 82]. In the latter one the authors additionally applied an algorithm called RePair, which works like the standard Sequitur, but in an offline fashion and thus with lower computation time. The authors of [25] also implemented it in an iterative framework to overcome its limitation of only finding identical matches. Prior to the current work, in [70] the Sequitur was already assessed regarding other pattern discovery approaches. It has the ability to run enormously fast and thus giving the user a quick feedback of what kind of patterns are within the data set. Additionally in [70] a pattern enumeration algorithm was proposed to use instead of Sequitur, due to Sequitur algorithm's limitation of ignoring some covered rules. In general, pattern enumeration is a word count algorithm, that enumerates the count of every pattern that is included in the time series in a certain range of pattern lengths.

Preprocessing Techniques

The discovery of patterns based on symbolic time series is highly influenced by the preprocessing techniques that perform the transformation from a real-valued time series to a symbolic one. These techniques reduce the effort of calculating a similarity within the actual process of pattern discovery. The definition of similarity therefore is included implicitly within these preprocessing techniques, which is why it is important to consider other techniques than only SAX. In [6] Azulay et al. additionally discussed other methods for discretization of time series:

- Knowledge-based discretization: User defined discretization boundaries.
- Equal width discretization: Partitioning of signal value range in equal sized bins, given a number of bins as well as the minimum and maximum values.
- Equal frequency discretization: Partitioning of signal value range into a given number of bins, having approximately the same frequency in each bin.
- k-means clustering: Assign all samples to k bins by applying the clustering algorithm.
- Persist: Taking into account the temporal structure of the time series, by maximizing the duration of intervals of equal state.

Beside this, the authors of [89] proposed an adaption of SAX, which takes into account variable sized windows. If a high variance in the data is detected within the current window, the window size can be shortened. In situations with low variance the window is adapted to be longer. A fundamentally different approach to time series discretization is proposed by Ritt and O'Leary in [78] by applying linear differential operators to the time series before discretizing it. Furthermore, a well known additional preprocessing technique is *numerosity reduction*, which is applied to the symbolic time series [52] [25] [70]. The method reduces every equal consecutive symbol in the time series to one of its kind. Hence, it enables the following algorithm to detect patterns that are scaled in time.

Multiresolution Discretization
Until now all the above described algorithms for discretization-based pattern discovery require the user to fix the number of bins. Because the result of pattern discovery is highly dependent on the number of bins and the chosen discretization technique, it is desired to reduce this effort of parametrization. In [85, 86] Shieh and Keogh introduced a multiresolution SAX algorithm called iSAX designed for time series indexing, which discretizes the time series with multiple number of bins. This time series representation was used in [10] as a basis for the developed motif discovery algorithm *MrMotif*. It is based on the idea of first processing the symbolic time series of low resolution and then expanding the result to higher resolutions. Hence, instead of fixing the number of bins or the boundaries itself a minimum and a maximum resolution is needed as an input.

Another algorithm was recently introduced by Gao and Lin in [24]. They adapted the iSAX algorithm in order to create a structure that is better accessible in the context of motif discovery, because iSAX in general is made for time series indexing. While iSAX always doubles its resolution, the proposed algorithm increases the number of bins incrementally. They additionally introduced a new data structure, the adaptive SAX Forest (aSAX-Forest), which is specifically made for grammar-based pattern discovery. Furthermore, they introduced the distance propagation Sequitur algorithm, which not only stores the hierarchical structure of the symbolic time series, but also statistical measures regarding every symbol. These statistical measures can be used for a distance propagation and subsequently as a threshold-based condition to form new rules.

Beyond Identical Sequence Matching

A major issue of Sequitur, which is the most used algorithm for discretization-based pattern discovery, is the necessity of identical symbolic matches. Two subsequences can only be assigned to the same pattern if their symbolic sequences are identical. Especially in the case of long subsequences with qualitatively similar shape the existence of noise or short outliers can cause the corresponding symbolic subsequences to be non-identical. This issue is addressed by multiple publications with different techniques, e.g. [7, 12, 25, 26]. For the sake of completeness, the random projection algorithm of [12] also adds the desired robustness, but has the limitation of fixed length pattern discovery. In [7] the authors increased the tolerance of Sequitur by allowing a rule to have digrams with varying symbols within a threshold based on the MINDIST measure. Therefore they extended the dictionary by a structure that stores all the assigned digrams and a calculated representative digram.

Another solution was proposed by Gao et al. in [25] by the iterative algorithm *itrSequitur*. This algorithm initially discovers patterns based on the standard Sequitur algorithm. Then the algorithm sorts out patterns that are not valid to subsequently reconstruct the initial symbolic time series with the valid set of patterns. Within the subsequent Noise Reduction Operator, symbolic subsequences, that resulted from noise and would prevent long patterns to be discovered, are deleted in the symbolic time series. This reduced symbolic time series can again be used as an input to Sequitur to start the next iteration. Doing so, it is more likely to discover long patterns in noisy time series. Recently Gao and Lin proposed another algorithm in [26] to address the same issue, which is based on a greedy strategy and improved by a recursive subfunction to additionally find non-identical subsequences.

Handling of Multivariate Time Series

In general, there are three different strategies to perform multivariate pattern discovery. The first one is to transform the multivariate time series into a univariate representation e.g. by applying Principle Component Analysis (PCA) like in [46, 90, 91]. Unfortunately, there is no guarantee that PCA or other techniques are able to reduce the dimensionality to one without significant loss. The authors of [18, 78] solved this problem in the domain of classification by discretizing all dimensions separately and then transforming them into one univariate symbolic representation by combinatorics. This univariate representation can simply be analyzed by univariate pattern discovery approaches. The second strategy is to adapt the pattern discovery algorithm itself. Minnen et al. [63] adapted the random projection algorithm to first calculate all-dimensional patterns. Additionally, they proposed a relevance estimation for every dimension of each pattern. Last but not least, the third possibility is to first apply the pattern discovery separately to each of the dimensions and then analyze the co-occurence of patterns in different dimensions. Vahdatpour et al. proposed in [95] the calculation of a coincidence graph based on the univariate results and the application of a graph clustering method. The authors of [8] introduced a neighborhood distance, which is basically the overlap of patterns in different dimensions, and grouped together the patterns regarding their distance and a threshold.

Postprocessing Techniques
Even though pattern discovery algorithms are made to give an overview regarding the patterns that can be found within the time series, most algorithms still output a high number of patterns, which is not transparent for the user. As evaluated in [70], this especially but not exclusively is a problem of discretization based approaches. Depending on the users background different kinds of pattern can be of interest. Hence, postprocessing techniques are necessary to e.g. rank the patterns regarding the users interest or their definition of relevance (e.g. [10, 57]). In most cases the relevance is a function of pattern-frequency, -complexity and -similarity. Beside this, some authors also mentioned the requirement of finding patterns that represent the time series [26]. The related formal problem of this challenge is called *set cover problem*. Informally, it can be translated into finding the smallest set of patterns that collectively cover the whole time series. Because the set cover problem itself is NP-hard, the authors of [3, 11, 22] proposed greedy algorithms for approximate solutions. Beyond this, Bashar and Li described in [9] another kind of postprocessing, i.e. the idea of automatically interpreting discovered patterns, and additionally reviewed approaches to reduce the diversity of patterns.

2.1.4   Other Approaches

Probably one of the best suited similarity measures for requirements like variable length and time scaled pattern discovery is Dynamic Time Warping (DTW). In general DTW is a distance or similarity measure for relatively small time series with different length with the possibility to stretch or compress one of the time series to fit the other one [37]. For two sequences T_1 and T_2 with length n_1 and n_2 DTW calculates an $n_1 \times n_2$ distance matrix by sample-wise comparison under consideration of distances of previous samples. The naive brute force approach to solve the pattern discovery by means of DTW is to use it like the Euclidean distance in the Matrix Profile. Unfortunately, this would still constrain the analysis to compare subsequences with equal length, even though it would now be able to identify time scaled patterns. Furthermore, the computational complexity of the brute force approach would increase even more because of the need to calculate a matrix for every subsequence-pair comparison. To reduce this amount of computational effort the authors of [76] did an extensive review concerning speedup techniques for DTW and additionally proposed a method that is said to be faster than 2012th state-of-the-art Euclidean distance calculations. Beside this, the authors of [94] proposed a one-pass algorithm called CrossMatch, which is able to discover patterns by once calculating the $n \times n$ distance matrix of the time series of length n with itself instead of pairwise comparison of every subsequences T_i and T_j. By small adaptions of the scoring function for each cell within the distance matrix, it is possible to detect subsequences of high similarity. A similar technique was proposed in [44] with an additional hierarchical clustering algorithm, which groups together subsequences that were previously extracted by the calculation of a distance matrix. For more details the reader is advised to read the prior work [70], which includes the test and evaluation of this one-pass DTW technique.

In 2007 the authors of [64] proposed a method based on a combination of k-nearest neighbor, density estimation and hidden Markov models (HMM). They first determine the k-nearest neighbors for every subsequence of length m in the time series by pairwise distance calculations. Subsequently for every subsequence and their k-nearest neighbors a density is calculated. Iteratively the highest density subsequences are extracted and modeled by the HMM to refind the shapes in the whole data set. Furthermore, they identified the advantages of using DTW within their distance calculations in order to find variable length and non-linear time scaled patterns.

In [99] Keogh et al. proposed a method based on the random projection algorithm from [12]. They used this algorithm to identify regions of interest for possible motifs and then applied the uniform scaling distance (USD) measure to extend these approximate solutions to variable length motifs. USD is able to perform the Euclidean distance calculation between two subsequences multiple times with different time scaling factors. In doing so, it is possible to compare subsequences with varying length. USD finds the best matching scaling factor between both sequences and outputs the distance of the best match. For motif or pattern discovery the scaling factor is limited. Hence, the patterns can have variable length, but only in a limited range from the initial subsequence length.

The authors of [88] proposed two approaches to reduce the effort of finding motifs by utilizing index structures. Both approaches are based on index structures (R*-tree and Skyline index) which store the approximate distance in a kind of hierarchical structure and, hence, can be used to quickly access the nearest neighbor information. The approximate distances are calculated by transforming the subsequences to a lower-dimensional space. This is done in the first approach by the creation of minimum bounding rectangles and a group distance function. The second approach builds the lower bounding distance by a measure called *Middle Points and Clipping* (MP_C). After finding candidates the exact distances are calculated by the Euclidean distance and an early abandoning technique.

As well as the previous approach, Torkamani and Lohweg [17] proposed an approach to transform the real-valued time series into a new feature space. Therefore they extracted subsequences with varying lengths by a sliding window and applied the *Quad Tree-Complex Wavelet Packet* (QT-CWP) to each of them. Hence, in the new feature space created by QT-CWP they are able to discover motifs and patterns of variable length and with shift-invariance-property.

In [30] the authors formulated the pattern discovery as an optimization problem of a function that depends on the frequency of k chosen patterns. They proposed a gradient ascent technique to maximize this objective function. Therefore, their algorithm randomly chooses k subsequences of length m at the beginning, which are then iteratively updated to maximize the objective function. Unsupervised pattern discovery approaches based on neural learning procedures can only rarely be found, which is why in prior work an artificial Neural Networks (NNs) was applied to this problem [72]. An approach was proposed that transforms the time series into an image and then uses a convolutional autoencoder to iteratively train the time series. It was shown that the autoencoder is able to encode the time series to a lower-dimensional space, whereas these lower dimensions refer to the shape of patterns and the corresponding latent spaces display the occurrence of the patterns.

Focusing on automotive data, the authors of [34] proposed the *Toeplitz Inverse Covariance-based Clustering* (TICC), a new method for multivariate time series clustering. While most approaches focus on matching the real-valued samples of the time series, the TICC analyzes the correlations of different sensor-signals within a window of varying length. Hence, for each cluster a correlation network is built based on a Markov random field method. Because this method does not take into account the actual shape of the time series, it can not be classified as a pattern discovery method in the sense of this work.

As many techniques were already described to reduce the effort of computation, the authors of [77] added another approach. They described the possibility to constrain the pattern discovery to interesting parts of the time series. By applying common pattern discovery algorithms only on these parts a major speedup can be expected. In this work the authors proposed a method to automatically filter the interesting parts by an algorithm called *Multidimensional Permutation Entropy*.

2.2 Representative Cycles

Representative cycles are widely used to represent the real-life usage of technical devices in a compressed way. The need for representative cycles directly derives from the variety of the real-life usage and the large amounts of data that is available from the device under surveillance. From an engineering perspective the exact knowledge of how the devices are used in the real life has big advantages regarding e.g. the design and control strategy of the device itself. Hence, methods to analyze the usage are of tremendous interest. The construction of representative cycles is one data-driven method to transform the large amounts of data to manageable sizes and therefore reduce the effort of testing or add a data-driven perspective to the engineering process.

Besides the engineering perspective, especially in automotive applications representative cycles also have a high relevance for the homologation of vehicles. In this case the cycles are used as tool for comparison in terms of energy consumption and emission of CO_2 (carbon dioxid), NO_x (nitrogen oxide) and particulate matter. According to the decision of the European Union the average new sold passenger and light duty commercial vehicle is not allowed to emit more than $95 \frac{g_{CO_2}}{km}$ from 2021 on, while this limit was recently proposed to being reduced even more [1]. Standardized test procedures based on representative cycles are therefore used to control these limits and eventually penalize the manufacturers or even prohibit to sell certain models, which exceed the limits of NO_x emissions. Because these cycles are always referring to the driving of those vehicles, the terms *driving cycle* and *representative driving cycle* are now defined in contrast to the basic term *cycle*:

Definition 5. *A driving cycle (DC) is a time series containing (at least) a signal, which describes the vehicle speed, and a corresponding sequence of time. Depending on the application it is possible that a DC also includes further signals, like the acceleration or the slope of the street. The length and the sampling frequency of a DC also varies with the application.*

Definition 6. *A representative driving cycle (RDC) is a DC, that is representative for an input data set regarding selected features, i.e. the feature values of the RDC are equal or at least similar to the feature values of the input data set.*

In the following sections we will take a closer look at possible applications for RDCs (Section 2.2.1), different definitions for representativeness (Section 2.2.2) and existing methods to construct RDCs (Section 2.2.3).

2.2.1   Possible Applications

In general, possible applications for RDCs can be divided into 3 different types. First of all, RDCs are widely accepted tools within the **engineering** process for design, testing and simulation, because of their ability to describe the real-life usage of the vehicle and the behavior of the customer. Hence, they are used for functionality and endurance testing in environmental conditions that are close to real life. This testing can be done in real life as well as in simulations or in hybrid solutions, like *Hardware* or *Software in the loop* test benches (HiL, SiL). This is an important step when optimizing the components and their control strategies. Additionally a common procedure in the automotive industry is to design a component for the *99 % customer*, which will lead to a component that is able to withstand the stress and the load in 99% of all cases. An RDC that captures a stress level higher than in 99% of all cases can be used for this purpose. Beyond this, RDCs are also tools for the comparison of different powertrain topologies or settings, see for example [75].

Secondly, they are used for the **homologation** process of vehicles to check if its emissions are within regulatory boundaries. For example the New European Driving Cycle (NEDC) has been used until 2017 for the standardized calculation of fuel and energy consumption by driving the NEDC with every new vehicle model on a chassis dynamometer. In 2017 the NEDC was replaced by the Worldwide Harmonized Light Vehicles Test Cycle (WLTC) coming along with stronger emission regularities and a new measurement procedure. While the WLTC is also driven on a chassis dynamometer, the legislature also established Real Driving Emission (RDE) regularities for the NO_x emissions that are allowed within trips of real driving ($126\frac{mg}{km}NO_x$ for gasoline engines, $168\frac{mg}{km}NO_x$ for diesel engines) [33]. To be suitable for the testing procedure, these RDE trips need to fulfill some requirements concerning e.g. velocity, acceleration and duration (for more details see [53]). Because of the increasing difficulty of ensuring the vehicles to emit less than these limits, Liessner et al. [53] proposed a methodology to construct RDCs that are within the RDE requirements. This enabled them to test their vehicles for RDE regularities before actually driving them.

Furthermore, RDCs are used for the analysis of **traffic** within limited spatial regions like cities. In general, their aim is to gain experience about the traffic condition in and around spatial regions of high traffic load, in order to reduce emissions and improve the air quality, see e.g. [56] [36]. Depending on the application the RDCs are varying regarding the chosen construction method, the length of the cycle, the included signals and the chosen features for representativeness.

2.2.2 Representativeness

Representativeness is the ability to make statements from a small sample of data concerning a much bigger amount of data or the real life in general [59]. Therefore we have to consider two kinds of relations when speaking about representativeness:

1. Relation between the real world and the logged data itself.
2. Relation between the logged data and the representative sample (RDC).

The first relation says, that the data needs to capture all the important information that is necessary for the application. Hence, the users have to clarify for example what information has to be captured and what level of detail is necessary. The second relation addresses the problem of how to compress the diversity of information within these large amounts of data to a short RDC. The RDC can only be as representative for the real world as the logged data is. If the user has not given enough attention to the first relation, the logged data will not be able to capture the important information and subsequently the RDC will not be able to do that, too. This work will focus on the second relation. All the problems concerning the first relation are assumed to be solved.

In terms of measurement of representativeness, there is an unspoken agreement in this research field to use knowledge-based features, that need to be defined beforehand and are customized to the application. Most of the publications, that are especially cited in Section 2.2.3, calculate statistical values (mean, median, standard-deviation, minimum, maximum) of velocity, acceleration, deceleration or other signals for the logged data as well as for the constructed RDC. The quality or representativeness of this RDC is then measured by calculating the error between them in the feature space. Regardless of which features are used for the measurement of representativeness, these features are generally called *Characteristic Parameters* (CP).

Beyond these statistical features, it is also possible to use the percentage of time spend in states like cruising, idling, acceleration and deceleration, which was proposed in [14]. Additionally, they used the *Vehicle Specific Power* (VSP), which is derived from the formula of driving resistances. The authors of [75] made small adaptions by only using the positive VSP, which is called *Mean Tractive Force*

(MTF). The MTF is closely related to the energy consumption, because it describes the power that is necessary for propulsion. Using the energy consumption itself is usually more complex since there is the need for simulation models that capture the efficiency of every component in every state. Of course, the energy consumption can also be measured in the vehicle itself, but this requires the RDC to be driven in the environmental conditions that are given from the real live usage. The energy or fuel consumption is a common feature for representativeness, as shown in e.g. [20, 39–41].

As we already discussed different purposes of RDCs in the previous section, an RDC can furthermore have different objectives depending on the application. First of all the RDC can be designed to represent the average of the input data in a certain feature space. This RDC will describe how the vehicle or the component is most probably used. But, especially in the context of design and engineering of components, it can be beneficial to construct a worst case RDC for the 99% customer. An RDC, that is constructed with this objective, would represent a stress level, which is higher than in 99% of all cases. Beside this application, the need for worst case RDCs is also mentioned in [53] in the context of RDE conformity.

Beside the consideration of average and worst case RDCs, many publications additionally differentiate between the consideration of the input data in total and the input data separated in trips. A trip is defined as a driving cycle between two consecutive system shutdowns of the vehicle. This small detail expresses itself in Section 2.2.3, where there is a need for the RDC to be an existing trip instead of a synthetically created one.

2.2.3 Existing Construction Methods

The varying purposes of RDCs spread a wide field of requirements concerning the construction methods and the cycles itself. Depending on the application and the characteristic parameters for representativeness, different construction methods can be considered.

Definition 7. *A construction method is a procedure to construct an RDC based on real-life and/or simulated data. This procedure can be described as the selection, adaption, synthesizing or semi-synthesizing of DCs.*

In the literature many different attempts to classify the construction methods can be found. For example [23] theoretically clustered the approaches regarding micro trips, segments, pattern classification and modal events. In [41] and [40] the authors classified them as micro trip, Markov chain and fuel based method. They additionally implemented these methods and compared their results concerning the representativeness. Because each of these contributions only considered parts of the big picture, this work proposes a slightly different way to classify the approaches. Based on extensive research the majority of construction methods can be divided into four different categories:

1. Micro trip-based methods.
2. Modal event-based methods.
3. Pure Markov chain-based methods.
4. Full trip methods.

These categories are summarized in Fig. 2.3. The figure shows a high level flowchart of a variety of functions and their sequence, which were used by other researchers and developers. In the following subsections the different approaches will be explained in more detail by referring to Fig. 2.3 and the corresponding function block numbers. However, there are still some methods that are highly specialized to the respective application, which is why the last subsection additionally treats other approaches like hybrid methods, that utilize multiple approaches.

Before going into detail concerning the different approaches, it is useful to introduce the concept of *candidate cycles*. Many methods construct multiple candidate cycles as an intermediate result, see block 8 in Fig. 2.3. Afterwords in block 9 these candidate cycles are evaluated regarding the chosen characteristic parameters for representativeness and, as part of block 10, the most representative ones are then chosen to be the representative cycles. In some applications it is preferred to construct just one RDC, while in others there is the need for multiple RDCs.

In order to construct multiple RDCs an alternative is to initially divide the input data set to multiple classes, like in [14]. They classified every data sample regarding its current speed, the VSP, breaking and idling. By applying a PCA and a clustering algorithm they identified 4 different classes of trips. For each of those classes they calculated an RDC.

Micro Trip-based Methods
The first category of construction methods are the micro trip-based approaches. They are widely accepted as common construction methods, especially in Asia, where they are used e.g. for regional cycles of Malaysia [62], Changchun [109], Kuala Lumpur [92] and Singapore [36]. To construct candidate cycles at first, they usually use the function blocks 2 (*Segmentation*), 4 (*Clustering of segments*) and 6 (*Chaining*) in Fig. 2.3.

A common technique to segment the real-life vehicle data into micro trips is to cut the data at zero (or nearly zero, e.g. [110]) vehicle speed, which is usually called "zero to zero" segmentation. These micro trips are then clustered into similar groups of micro trips according to some predefined features and a selected clustering algorithm. In the following step usually the centers of the micro trip clusters are selected and then randomly chained together until the candidate cycle exceeds a predefined length of the cycle. The process of chaining in this case is simple, because of the condition of zero velocity at both ends and the RDC being univariate. After creating multiple candidate cycles, the representativeness of each cycle is calculated (block 9). On this basis one or more cycles are selected to be representative (block 10).

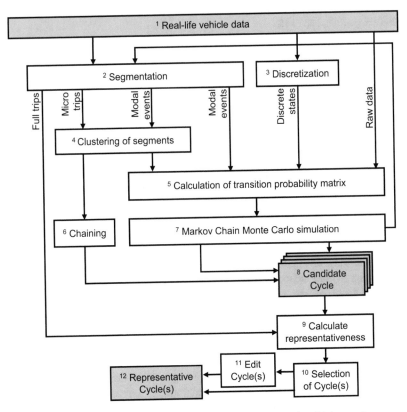

Figure 2.3 Overview of existing methods to construct representative driving cycles

While this process is the most common one, regarding the micro trip based approaches, there is still enough space for a lot of variety within the applied functions. For example, the clustering of micro trips is mostly done by defining some knowledge-based features, as already shown in Section 2.2.2. For clustering the authors of [111] used a method combining a k-means clustering with a Support Vector Machine (SVM). A dynamic clustering algorithm preceded by PCA was proposed in [109].

In [83] a special chaining methodology was proposed. They constructed one candidate cycle by randomly concatenating micro trips. But instead of generating multiple candidate cycles, the one initial cycle is enhanced by replacing single micro trips with others that will push the values of the characteristic parameters closer to the average. In this manner the candidate cycle will get more and more representative with every iteration. Translated to the flowchart of Fig. 2.3, they added the block 11 and skipped the block 10.

While chaining is simple in the case of "zero to zero" segmentation, this step gets more challenging when the segmentation is done "zero to non zero" or vice versa. This problem is addressed in [5, 62] and is closely related to the modal event-based construction methods. The authors of [62] implemented a segmentation with fixed length, which is why there will eventually be micro trips that do not start or end with a vehicle speed of zero. They additionally proposed two chaining conditions of micro trips, to not include abrupt jumps in the velocity of the RDC. First, to concatenate two micro trips, the velocity needs to be in a $1 \; km/h$ range. Secondly, the "condition" of the vehicle has to be the same in the chained parts of both micro trips. The condition corresponds to the state of acceleration of the vehicle and is roughly classified in acceleration, deceleration and cruising.

In the ARTEMIS project (assessment and reliability of transport emission models and inventory systems, [5]) the authors identified 12 typical driving situations (e.g. *congested urban with high stop duration*) by segmenting the data to micro trips of equal size and clustering it on the basis of a speed-acceleration distribution of every micro trip. In comparison to [62] for the chaining of micro trips they additionally included the transition probability of the vehicle going from one situation to another.

In general after selecting an RDC there is the possibility to edit the cycle to enhance its representativeness (see block 11). In [80] the authors included a representative amount of idling time between the micro trips. Similar to this, in [83] the proposal of additional parking time was made.

There are also some special methods for segmenting the data. For example in [79] the authors focused on creating an RDC for a certain route, which is why they segmented the data based on road segments and infrastructural information.

Modal Event-based Methods

A modal event is a short, physically meaningful piece of the time series. In the context of RDCs it was first described in 2002 by the authors of [56], where they referred to it as small scale acceleration or deceleration modal events. In theory a modal event can have a multivariate definition, but in most publications it is reduced to the speed of the vehicle. The extraction of modal events from time series data can be classified as a segmentation method (block 2), while in most methods the length of a modal event is not fixed, but can vary. In [14] the authors described that the modal events duration can vary between approximately 1 and 100 seconds. Obviously this definition of a modal event comes close to our definition of a pattern in time series.

In general, when applying a modal event-based method for cycle construction, the same function blocks like in the micro trip-based approaches can be used, namely block 2 (*Segmentation*), 4 (*Clustering of segments*) and 6 (*Chaining*). For segmentation and clustering the authors of [55, 56] utilized a maximum likelihood estimation (MLE) partitioning algorithm, that clusters segments of varying length and assigns a specific modal operation condition.

Dai et al. proposed in [14] a rule-based segmentation method, which at first classifies sequences of varying length to four different driving conditions acceleration, deceleration, cruising and idling. This method is parametrizable and was optimized by a weighted sum-based comprehensive assessment parameter (CAP), which was also proposed in this paper. These driving conditions are then divided again into four subgroups, based on the average speed, for final definition of modal events.

Within the modal event-based methods a widely accepted approach is to replace the chaining by a combination of block 5 (*Calculation of transition probability matrix*) and block 7 (*Markov chain Monte Carlo simulation*) [14, 55, 56, 97]. At first a transition probability matrix (TPM) between all modal events is calculated based on the input vehicle data. A transition probability $P_{A,B}$ describes the probability of the vehicle experiencing consecutively modal event A followed by the modal event B. The matrix containing all the transition probabilities from every modal event to every other modal event is then input to the Markov Chain Monte Carlo (MCMC) simulation. The n-th order Markov chain describes the state of a system being dependent on the n states before. The first order Markov chain is also applicable to modal events. Beginning with an initial modal event, a candidate DC can be iteratively synthesized by choosing the next modal event based on the transition probability matrix and a randomized experiment. This procedure is inspired by the Metropolis-Hastings algorithm to construct a sequence of states. The simulation is done, when an abort criterion is met. Usually this criterion is defined as the exceedance of the preferred length of a DC and the ending of the current modal event in a vehicle speed of zero.

Pure Markov Chain-based Methods

As described in the previous section, the usage of Markov chain theory is a widely accepted technique to concatenate predefined states in a statistically valid manner. However, states do not need to be defined as modal events. Within the pure Markov chain based methods these states are usually only discretized values of a time series, describing in most cases the velocity and the acceleration of a vehicle. Referring to Fig. 2.3 we are speaking of function blocks 3 (*Discretization*), 5 (*Calculation of transition probability matrix*) and 7 (*Markov Chain Monte Carlo simulation*).

It is conspicuous that there is also a connection between block 1 (*Real-life vehicle data*) and block 5, describing the raw data path into the calculation of the TPM. Even though a few authors are referring to this path [29, 50], this is, strictly speaking, not correct, because the TPM calculation always demands discrete states. In their case they logged the vehicle data with a low accuracy, e.g. 1 km/h for velocity and 0.2 m/s^2 for acceleration, which is why a discretization was not necessary.

Lee et al. proposed and applied the pure Markov chain-based approach in multiple publications [49–51]. They calculated the TPM regarding two dimensions, velocity and acceleration, and then created multiple candidate DCs by applying the MCMC simulation. The simulation stops, when the length of the created DC exceeds some predefined distance threshold and the values of 8 important statistical parameters match approximately the parameter values of the input vehicle data. These 8 statistical parameters were previously chosen from a set of 27 parameters by regression analysis. The authors of [29] applied the same method, except for the abort criteria of the MCMC simulation, which they simply defined as exceeding a distance of 14.1 km and ending in zero velocity.

Nyberg et al. utilized the pure Markov chain-based approach in [75] and proposed some adaptions to enhance the comparability of RDCs within the engineering precision. They experienced some implausible acceleration values in high velocity regions, which is why they added an acceleration constraint in the MCMC simulation. Furthermore they proposed three performance measures α, β and γ related to the aerodynamic drag, rolling friction and intertia force driving resistances. Hence, two DCs are equivalent if they have the same values for α, β and γ. To push the equivalence of the resulting candidate DCs even more they proposed to edit them by slightly altering the samples within positive traction phases, especially at the beginning and ending of those phases, and by expanding or contracting the time series while keeping the average speed constant.

As mentioned earlier, it is also possible to use higher order MCMC simulation, which is shown in [21]. In this publication the authors required the RDC to represent additionally the slope of the street. To overcome the difficulty that an increase in velocity and a change in slope needs to be analogous to the acceleration, they chose a 3rd order Markov chain. Furthermore, for computational efficiency they calculated a TPM only for the states that do occur in the real life.

In general, when applying the MCMC simulation to construct RDCs, the user can not be sure if or when the characteristic parameters for representativeness will converge. This is because of the random sampling in the simulation. Usually, this problem is faced by creating multiple candidate DCs and then choosing the most representative one. An alternative solution is presented in [108], which combines the traditional Markov chain approach with a genetic algorithm (GA) to accelerate the convergence and to improve computational efficiency.

Full Trip Methods
In comparison to the other approaches for the construction of RDCs, the full trip methods are in general much easier. They eliminate the effort of chaining segments by choosing existing trips to be the RDCs regarding some predefined features. In general they choose the trip closest to the center in the feature space. Doing so, these approaches only need a simple segmentation method (block 2), to define starting and ending points of trips, and a calculator for characteristic parameters (block 9). Because these RDCs are not synthesized, this construction method is also classified as deterministic, while all the other methods are stochastic.

The authors of [39, 40] proposed a fuel-based method, where they calculated the energy consumption of a certain vehicle for every trip and then chose the trip having the energy consumption closest to average. Tzeng and Chen went a step further in [32] and calculated 11 features describing the dynamics of each trip. By applying a factor analysis method, they reduced the dimensionality of the feature space and prevented the analysis from bias. An additional postprocessing was proposed in [83]. After calculating 25 features and choosing the center trip they enhance the trip by replacing single micro trips until the error falls below a threshold. Beyond this, they add representative parking time to the RDC in order to construct a 24 hour cycle.

In [107] the researchers described a procedure of analyzing similarities of existing RDCs like NEDC (New European Drive Cycle), JC08 (Japanese chassis dynamometer test cycle from 2008) and FTP72 (US Federal Test Procedure) in order to reduce the effort of testing to a small set of RDCs instead of testing every single existing RDC. Therefore they proposed two methods based on the analysis of over 60 statistical features and one additional method, which is based on the Euclidean distance calculation of the RDCs' speed profiles.

Hybrid and Other Methods

Beside the four approaches, that were described previously, there are further methods, which can not be uniquely classified due to e.g. their high degree of customization. For example in [106] the authors chose a simple approach to combine Markov chain- with micro trip-based methods. They first extracted and clustered the data into similar micro trips and then calculated a TPM between these clusters and did an MCMC simulation.

The combination of Markov chain- and micro trip-based methods can also be implemented the other way around, as proposed in [19, 20, 53]. By calculating the TPM on board of the vehicle based on raw or discretized data, it is possible to preserve the driving characteristics of the vehicle and the driver without logging all the data. The TPM can afterwords be used to do MCMC simulation and to synthesize a data set that has similar characteristics of the original data. In [53] the authors built one-dimensional RDCs by chaining synthesized micro trips in a way to maximize vehicle dynamics and to meet RDE regulations like a minimum duration, as well as a certain distribution of driving modes. The authors of [19, 20] focused on the multivariate RDC construction. In order to chain multivariate micro trips, they synthesized these micro trips with the boundary condition of starting and ending in reference states, which in general are the most frequent states in the database. Because of this boundary condition, the chaining even in higher-dimensional spaces gets much easier.

A highly customized method was proposed by Hongwen et al. in [38]. The goal was to develop a cycle construction method, that runs in real time on board of the vehicle, in order to predict the range and fuel consumption. Therefore they used a Markov chain approach and merged it with a real time traffic information tensor.

2.3   The Idea of Pattern-based Cycles

This sections sketches the idea of a *pattern-based RDC construction* approach utilizing the unsupervised pattern discovery of Chapter 3 as a segmenting method. While existing RDC construction approaches of Section 2.2.3 mostly use highly customized segmenting methods, the unsupervised pattern discovery is developed to be applied to various use cases. The general idea of the pattern-based RDC construction is to replace the segmenting methods by a powerful pattern discovery method, that is able to process large amounts of potentially multivariate time series data in a highly automated manner without the need of extensive expert knowledge. To use it effectively for RDC construction the discovered patterns need to be able to represent the majority of the input data, meaning the set of patterns and all their members should cover almost any sample in the data.

Based on the results of the pattern discovery it is possible to characterize a cycle by the patterns, that were discovered in it, and their frequency. The basic assumption here is, that the patterns that were discovered in the input data, should also occur in the RDC. In fact they should not only occur once in the RDC, but as frequent as in the input data related to its total length. Hence, the relative coverage of a pattern in the input data should be approximately the same as the relative coverage of this pattern in the RDC. For both, the input data set and the RDC, it is now possible to calculate a relative coverage distribution. Both distributions are characterizing their underlying data set. The error between these distributions can then be used as a measure for the representativeness of a cycle. Using these unsupervised generated features it is possible to evaluate the quality without predefining any expert-based features as described in Section 2.2.2.

Following this pattern-based approach it is possible to implement two kinds of RDC construction methods. The first one is closely related to the full trip methods of Section 2.2.3, i.e. the calculation of a relative coverage distribution for every trip and subsequently the selection of the trip, which is closest to the center. The second possibility is the chaining of patterns in a reasonable sequence. In Chapter 4 these concepts will be explained in more detail.

Development of Pattern Discovery Algorithms for Automotive Time Series

3

This chapter deals with the development and optimization of algorithms for unsupervised pattern discovery in time series. While this work focuses on automotive time series data, the proposed approach, in general, is able to improve the handling of highly dynamic sensor data independently of its origin. As described previously in Section 2.1 there are many approaches that solve the problem of pattern discovery in time series. In previous work [70] the discretization-based approaches have already been identified as suitable for automotive requirements. This suitability is additionally discussed in more detail in Section 3.1 preceded by the explanation of the term *automotive time series*, its special characteristics and consequently the requirements for a pattern discovery method. Subsequently, a new framework for discretization-based pattern discovery in automotive time series is proposed in Sections 3.2 to 3.5. An overview of the framework and its features is given in Section 3.2. In Section 3.3 the reduction of complexity is described, which includes the discretization and further methods to subsequently improve the discovery of patterns. These extensions were already proposed in previous work [71] and are now included in the context of the whole framework. Section 3.4 describes the process of extracting a dictionary of patterns based on discretized time series. Both algorithms were already introduced in previous publications [70, 72] and are now again described in more detail. Finally, in Section 3.5 a method for the selection of relevant patterns is proposed.

Supplementary Information The online version contains supplementary material available at https://doi.org/10.1007/978-3-658-36336-9_3.

3.1   Suitability of Approaches

Pattern discovery in time series is a widely accepted research field and has been investigated in many different disciplines like seismology, medicine or music. Unfortunately, an application to automotive time series data has only been investigated to a small extend in [34]. Automotive data in this context means data recorded from vehicles, which includes sensor data as well as other time-dependent information generated by control units within the vehicle. In Section 3.1.1 special characteristics of automotive time series are investigated in comparison to time series recorded from other devices. Based on the derived requirements, the suitability of different pattern discovery approaches is discussed in Section 3.1.2.

3.1.1   Special Requirements in Automotive Time Series

In Section 2.1 the intuition of a pattern and its variance depending on the application was already addressed. This section describes the automotive intuition in contrast to other applications. Note that most of the described examples are based on speed and acceleration of a vehicle, while these examples can also be translated to other sensor signals, like torque, current, voltage or steering angle.

One major difference of automotive data in comparison to most other time series data is its high dynamics and diversity. In the medical context, the heart beat is often used as an example for the application of pattern discovery. In this case there is only a small number of different patterns to be found, namely the heart beat itself or different variants of heart beats. In seismological data most of the samples show only constant behavior until a seismological activity starts. Hence, it is also about finding rare patterns. Another example is data that is recorded from the usage of robots e.g. in production. These robots are always doing the same movements with a high precision. This will lead to patterns describing these movements, while the members of a pattern will probably have a high similarity. In contrast to these examples, vehicle time series data describing e.g. the speed or acceleration of a vehicle has much more diversity. This is because of the driver and the traffic environment. Vehicle data is majorly influenced by the behavior of the driver acting with or reacting to other road users or the infrastructure. Beside the pure variance of roads and traffic situations, every driver is behaving in a different way. That results in time series data that is equipped with many different patterns. Because of this diversity there is not the one relevant pattern, like a heartbeat, but a set of patterns that are relevant to describe the behavior of the time series. In terms of RDCs there is especially the need for finding all the patterns that are in total able to describe the majority of the time series.

Members of a pattern are required to be similar not just in shape, but also in value range. Hence, a pattern needs to be invariant to offset and amplitudinal scaling (see Table 2.1 b and c). This is because of the comparability of situations. For example an acceleration event from 0 to 30 km/h is not comparable to one from 0 to 60 km/h or even from 30 to 60 km/h, even though they might have a similar shape. The members within a pattern should include comparable situations. In contrast to this value range invariance, the patterns might be variant to a certain amount of longitudinal scaling (see Table 2.1 d). Imagine one acceleration event from 0 to 30 km/h with a two seconds duration and one with a three seconds duration. These situations may still belong to the same pattern, while an acceleration event with 10 seconds will probably be a different situation. Additionally, the pattern discovery has to be robust against noise, because no sensor signal is perfect and always includes a certain amount of noise (see Table 2.1 a).

Beside longitudinal scaling, the pattern discovery is required to discover patterns of variable length. While the longitudinal scaling describes the scaling within a pattern, the variable length property focuses on the varying length of different patterns. For example the method needs to discover an acceleration event of approximately 3 seconds, a deceleration event of 10 seconds as well as an approximately constant

Table 3.1 Comparison of different pattern discovery approaches regarding properties that are important for the applicability to automotive time series. The assessment of discretization-based approaches in this table is premised on the propositions of this work. The assessment of NN-based approach is premised on [72]. The assessment of DTW-based approach is premised on the implementation that was already tested and evaluated in previous work [70]. For the assessment of Matrix Profile approaches each publication regarding the UCR Matrix Profile was considered

Property	Matrix Profile	DTW	NN	Discret.
Invariance regarding offset and amplitudinal scaling	(✓)	✓	✓	✓
Longitudinal scaling (time warping)	✗	✓	(✓)	✓
Robustness against noise	✓	✓	✓	✓
Variable length pattern discovery	✓	✓	(✓)	✓
Multivariate analysis	✓	✓	(✓)	✓
Determinism	(✓)	✓	✗	✓
Effort of parametrization	low	medium	high	medium
Time and space complexity	high	high	medium	low

high-speed event of 2 minutes. It would be unsuitable to only find patterns with fixed length.

An additional requirement to a pattern discovery method that is suitable for automotive time series is the ability to process large time series. Imagine a vehicle with a mileage of 100,000 km and an average speed of 50 km/h. With a sampling rate of only 10 Hz the number of data samples of one signal for the whole vehicle life would count 72 millions. With higher sampling rates and/or greater mileage the time series length can easily exceed 500 millions of samples. Beyond that, automotive time series are enormously high-dimensional. In a vehicle a few thousand different signals can be recorded. However, for most applications only a small subset (< 10) of these signals is relevant. Even though guided, constrained as well as unconstrained search in multivariate time series is interesting for certain applications in the automotive context, this work reduces its investigation to the simple case of finding k-dimensional patterns in d dimensions with $k = d$. Furthermore the discovery of simultaneously occurring k-dimensional patterns is sufficient for most of the automotive applications.

3.1.2   Discussion of Suitability

Under consideration of the requirements collected previously, in this section a brief discussion of the suitability of different approaches described in Section 2.1 follows. As already showed in [70] and [72] the discretization-based approach has several advantages over other approaches based on Matrix Profile, DTW or NNs. In Table 3.1 the requirements as well as an assessment of each approach is listed.

Probably the biggest disadvantage of Matrix Profile is the focus on discovering patterns being variant to offset and amplitudinal scaling due to z-normalization of subsequences. Even though it is possible to replace the z-normalized Euclidean distance with a standard Euclidean distance, all of the proposed optimizations are based on z-normalization and are not necessarily transferable to standard Euclidean distance. Because of the usage of Euclidean distance the longitudinal scaling property is usually not fulfilled. On the plus side the Matrix Profile framework offers a variety of solutions to discover variable length patterns in multivariate spaces with robustness against noise. In general, the computational complexity of this approach is extremely high, as we already showed in [70]. But the authors proposed multiple techniques to speed up the computation for example by parallelizing the calculations, by preventing the algorithm to calculate the same thing multiple times or by offering a good approximate instead of the exact solution. The latter technique is the reason why determinism is not fully given. If the exact solution is calculated, the solution is deterministic. In case of an approximate solution the algorithm runs through the calculations in a random order, which is why the approximate solution can vary in two identical runs. The authors always aim for a parameter-free algorithm. Even though in most cases this is not possible, they succeeded in keeping the effort of parametrization at a low level. Unfortunately, Matrix Profile is rather used to identify meaningful rare patterns instead of a representative set of patterns. Nevertheless, it is a powerful framework because it offers solutions to many problems regarding time series analysis in general.

Pattern discovery methods based on DTW are best suited regarding the similarity requirements as shown in Table 3.1. The ability to *time warp* sequences, which includes longitudinal scaling and entails robustness against outliers and noise, is unique in pattern discovery and is highly desirable. Beyond this, these methods are able to deterministically detect patterns with variable length in multivariate spaces. Unfortunately, the computational complexity in time and space is too high to be applied to long time series on a regular basis.

Even though there is a lot of attention for NNs and their application in any domain, pattern discovery has only been investigated to a small extend. As shown in [72], by the use of convolutional autoencoders it is possible to discover meaningful patterns in acceptable computing time. The approach can be extended to do variable length pattern discovery on multivariate time series. On the downside there is an enormously high effort of parametrization for every new kind of data set. Additionally, it has the problem of non-determinism. Hence, with the same parameters executing on the same data set it is not guaranteed to find the same patterns due to e.g. random initialization. Both of the aforementioned problems lead to the conclusion that NN-based approaches are currently not suitable for automotive use cases.

Approaches based on discretization of time series have the disadvantage of being dependent on the number and location of discretization boundaries. Hence, the discovered patterns and their level of detail is majorly influenced by the users and their choice regarding the discretization boundaries. On the other side, these approaches in general have the benefit of fast execution due to reduction of complexity. The reduction technique in combination with text mining algorithms makes it the fastest approach for pattern discovery as already shown in [70]. Additionally, it enables the discovery of patterns with variable length under consideration of longitudinal scaling. These are the reasons to choose a discretization-based approach as a basis for further investigations regarding the pattern discovery in automotive time series. In the following sections extensions to the basic discretization approach are proposed to solve several problems and subsequently fulfill all the requirements listed in Table 3.1.

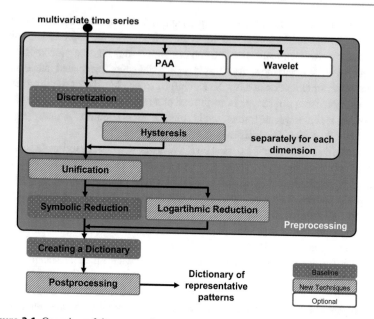

Figure 3.1 Overview of the pattern discovery toolchain including new techniques (marked as dashed boxes). Functions of the baseline method from previous work [70] are shown as dotted boxes. The clear boxes show additional and optional filtering algorithms [1]

3.2   Framework Introduction

Before going into detail concerning additional techniques to improve discretization-based pattern discovery in automotive time series, this section gives an introduction to the framework and the methods that are used as a basis. Fig. 3.1 shows an overview of the pattern discovery tool chain including all new techniques, which are marked as dashed boxes. The baseline methods, that have already been proposed, tested and evaluated in previous work [70, 72], can be identified as dotted boxes. In general the framework consists of three parts: preprocessing, creation of a dictionary

[1] "Intelligent Data Analysis, vol. 25, no. 5, Noering et al., Improving Discretization Based Pattern Discovery for Multivariate Time Series by Additional Preprocessing, pp. 1051–1072, (2021)", with permission from IOS Press. The publication is available at IOS Press through http://dx.doi.org/10.3233/IDA-205329.

and postprocessing. In Section 3.3 problems are described, that go along with the reduction of complexity. Beyond this, solutions are proposed that aim to solve these problems by adapting the complexity reduction process. This process includes all the techniques in the dark gray box in Fig. 3.1. Based on this compressed time series, text mining techniques are able to create a dictionary of discovered patterns, which is described in Section 3.4. This step marks the core of pattern discovery. Usually, text mining algorithms output a large amount of patterns, which is why a postprocessing step is necessary to select relevant patterns. Possible techniques to reduce the amount of patterns are introduced in Section 3.5.

The tool chain, that is used as a basis, includes algorithms for discretization, symbolic reduction and the creation of a pattern dictionary. Later in Chapter 5 it will be referred to as *baseline method*. Note that this tool chain is only capable of discovering patterns in univariate time series. The problem of multidimensionality is addressed in Section 3.3.2. Until then only univariate time series are addressed. In this work the simple approach for equidistant discretization of univariate time series is used. Unlike other approaches that use e.g. SAX for discretization, which includes averaging by PAA, the baseline method contains no filtering before the discretization step. The discretization transforms the real valued time series sample by sample to a sequence of symbols according to selected equidistant discretization boundaries, which result from a given number of bins *noSteps*. Symbolic reduction, also known as numerosity reduction technique, deletes all but one symbol in each consecutive sequence of identical symbols. This enables the subsequent method to discover longitudinal scaled patterns. By applying the *adapted Sequitur* or the *pattern enumeration* algorithm (see Section 3.4) it is possible to discover variable length patterns in univariate time series. A postprocessing step to select relevant patterns is not included in the baseline method.

3.3 Preprocessing—Reduction of Complexity

This section was taken from previous work [71] to a large extend. It describes the first part of the framework for discretization-based pattern discovery. The reduction of complexity always goes along with a loss of information. Typical problems and common solutions are introduced. Under consideration of the requirements formulated in Section 3.1.1 new techniques are proposed to face the issues of common solutions.

3.3.1 Problem of Symbolic Toggling

One of the biggest problems in discretizing time series is the symbolic toggling that results from noisy data, which is visualized in Fig. 3.2. The left side shows a subsequence of a univariate time series, which is classified into the symbols A and B by a specified discretization boundary. The resulting symbolic time series is displayed below the axes. Obviously there is a time interval with symbols jumping alternatingly between A and B, which is called symbolic toggling. This interval is highlighted in this and following figures as a gray box. Because of these irregularities, caused by the toggling, the patterns can not be identified in a trustworthy way. This is why a method is needed to reduce this toggling.

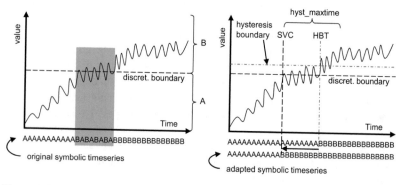

Figure 3.2 Problem of symbolic toggling (left) and solution by hysteresis (right)[2]

Common solutions are filtering methods to reduce the noise of the time series before discretizing it. One possibility is to apply PAA, which averages the time series depending on the parameters $windowsize$ and $stepsize$ and, hence, is able to reduce the toggling. On the downside PAA smooths the dynamic processes, which ends up being a limitation for the discovery of highly dynamic patterns. In the automotive context especially the dynamic patterns are of interest for the engineers, which is why these limitations should be avoided. Nevertheless, it still can be a powerful filtering method to reduce the symbolic toggling when the parameters are

---

[2] "Intelligent Data Analysis, vol. 25, no. 5, Noering et al., Improving Discretization Based Pattern Discovery for Multivariate Time Series by Additional Preprocessing, pp. 1051-1072, (2021)", with permission from IOS Press. The publication is available at IOS Press through http://dx.doi.org/10.3233/IDA-205329.

selected carefully. The issue of smoothing the dynamic processes can be solved by PLA. However, when using PLA the parameters also have to be set carefully to have the right amount of smoothing while preserving the dynamic processes. Furthermore, implying that the input time series is interpolated at equidistant time steps, it destroys the regular temporal structure of the time series. Note that this kind of temporal structure is helpful for the quality of the patterns due to the inability of the pattern discovery algorithms to take into account the time explicitly. The temporal dimension is handled more detailed in 3.3.3. To resume, what is needed is a filtering method which preserves the temporal structure and denoises the time series. Therefore, a frequency filter based on FFT or wavelet denoising could be used. Within the parameter experiments of Chapter 5 the wavelet denoising on different levels ($denoiselevel$) and the PAA with varying $windowsize$ and $stepsize$ will be tested.

Beside these common filtering approaches, now another solution to the problem of symbolic toggling is introduced. The concept, that is visualized on the right side of Fig. 3.2, is known as hysteresis or, mathematically speaking, backward bifurcation. Hysteresis is basically the delayed effect of a system after a change in the cause and is widely used to describe the physical behavior of e.g. the magnetization of iron. While in most cases the hysteresis can only be observed, we can use this concept to solve the problem of symbolic toggling. Therefore, we assume we already have a discretized time series $T_{disc} = d_1, \ldots, d_n$ with every d_i being the ith discretized value of the original time series T. To define a region of hysteresis we need two additional parameters. The first one $hyst_{value}$ describes the height of the hysteresis and states a relative value to define the $Hysteresis Boundary$ as:

$$HysteresisBoundary = DiscretizationBoundary \pm hyst_{value} \times h_{jumpinto}$$
$$(3.1)$$

The variable $h_{jumpinto}$ is defined as the height of the discretization interval that the symbolic value jumps into. The second parameter $hyst_{maxtime}$ defines the maximum duration of a hysteresis, which is necessary in case the value once exceeds the discretization boundary, but never breaks through the hysteresis boundary and instead stays in the hysteresis region for a long time. The algorithm runs through all possible symbolic value changes (SVC) chronologically and checks if it is a valid SVC. It is valid if the value of the original time series breaks through the hysteresis boundary within $hyst_{maxtime}$ time steps. The index of the hysteresis breakthrough (HBT) is named as hbt with $T_{disc}(hbt) = d_{hbt}$ and correspondingly svc with $T_{disc}(svc) = d_{svc}$. If it does not break through, it still can be a valid SVC if the value stays within the hysteresis interval for at least $hyst_{maxtime}$ time steps. In this case $hbt = svc + hyst_{maxtime}$ is the temporal HBT. Afterwards the symbolic values are adapted:

$$\begin{cases} T_{disc}([svc, \dots, hbt]) = T_{disc}(hbt) & \text{if SVC is valid} \\ T_{disc}(svc) = T_{disc}(svc - 1) & \text{if SVC is not valid} \end{cases} \quad (3.2)$$

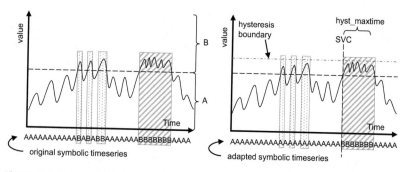

Figure 3.3 Behavior of hysteresis algorithm if the signal value never breaks through the hysteresis boundary. The figure also includes the special case if the value stays in the hysteresis region for at least $hyst_{maxtime}$ steps [3]

In the example of Fig. 3.2 the algorithm detects an HBT within the maximum duration when checking the first SVC from A to B. Therefore, the SVC is valid and the symbolic values from index svc to index hbt are changed to the value B. Thus the symbolic toggling is eliminated within the interval $[svc, \dots, hbt]$. Afterwards all eliminated unnecessary jumps do not need to be checked anymore.

The other two cases are sketched in Fig. 3.3. The dashed box shows a case with no HBT, but instead the duration of value B exceeds the maximum duration given by $hyst_{maxtime}$. Hence, it is still a valid SVC and the symbolic values will not be changed according to the algorithm. In the events shown in the dotted boxes the symbolic values need to be changed from B to A, because there is neither a HBT nor was the $hyst_{maxtime}$ exceeded.

[3] "Intelligent Data Analysis, vol. 25, no. 5, Noering et al., Improving Discretization Based Pattern Discovery for Multivariate Time Series by Additional Preprocessing, pp. 1051-1072, (2021)", with permission from IOS Press. The publication is available at IOS Press through http://dx.doi.org/10.3233/IDA-205329.

3.3.2 Problem of Multidimensionality

After proposing solutions to the problem of symbolic toggling, options for the multivariate pattern discovery are now discussed. As already mentioned, there are basically two approaches in terms of using the pattern discovery algorithms. The first one is to run the pattern discovery separately on each dimension and then analyze the sequence of the patterns in different dimensions. The second approach is to somehow reduce the dimensionality of the time series to one dimension and then run the pattern discovery just once. The first case is interesting because it opens up the possibility to find not only all-dimensional patterns, like in the second approach, but also shifted in time. Hence, it solves a more general problem than the second approach but also requires much more computation power. Due to the computing time optimization this work focuses on the second approach.

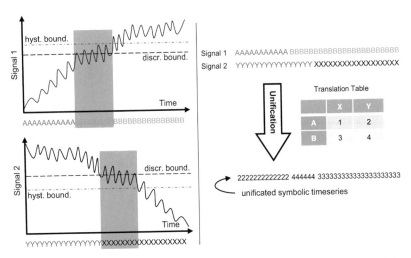

Figure 3.4 Examplary unification of two-dimensional symbolic time series by translation table [4]

---

[4] "Intelligent Data Analysis, vol. 25, no. 5, Noering et al., Improving Discretization Based Pattern Discovery for Multivariate Time Series by Additional Preprocessing, pp. 1051-1072, (2021)", with permission from IOS Press. The publication is available at IOS Press through http://dx.doi.org/10.3233/IDA-205329.

Dealing with dimensionality reduction, PCA is usually a good choice. For highly correlated multivariate time series, the reduction by PCA to one dimension will still preserve most of the information contained in the multivariate time series. But the assumption of high correlations between all time series seems to be unrealistic. Hence, even if it is possible to reduce the dimensionality by PCA to a few dimensions, we need a method to compress these to just one univariate time series, which we call *unification*. The authors of [18] and [78] had these problems in the domains classification and symbolic analysis of time series and already proposed similar solutions. Input for the unification are the discretized and optionally filtered (e.g. by hysteresis) time series of each dimension. A more detailed example for a two-dimensional subsequence is shown in Fig. 3.4. The left hand side shows the discretization process including the hysteresis of signal 1 (top) and signal 2 (bottom) separately. On the right hand side both symbolic time series are unified by transforming them into a new discretization space. This is done by calculating the Cartesian product of all symbolic vectors D_i of every dimension i containing the symbols s_{ij}:

$$D_{unified} = D_1 \times \cdots \times D_m \text{ with } D_i = \begin{pmatrix} s_{i1} \\ \vdots \\ s_{in_i} \end{pmatrix} \qquad (3.3)$$

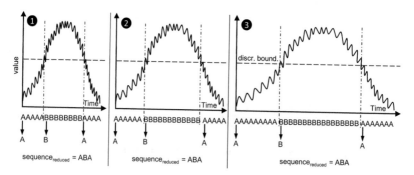

Figure 3.5 Example for the standard symbolic reduction with index relation

Note that every dimension i can have a different number of bins n_i. In doing so, a new symbol can be assigned to every element of $D_{unified}$. These assignment rules can be visualized as a translation table like in Fig. 3.4. In this example the switch from letters to numbers shows the change of the discretization space. By the proposed unification method it is now possible to transform any multivariate symbolic time series to a univariate symbolic time series. Nevertheless, it is recommended to keep the dimensionality low, due to the exponential growth of the length of $D_{unified}$.

3.3.3   Problem of Time Warping

As already shown in [70], a numerosity reduction technique, which is called symbolic reduction, is necessary in order to identify time warped patterns. Fig. 3.5 shows an example of the baseline technique with three different sequences being symbolically reduced to the same short symbolic sequence ABA. To maintain the information about the relation between reduced symbolic sequence and corresponding time, it is necessary to save the index values corresponding to the new reduced symbols. In case of this symbolic reduction, the new symbol always relates to the first index of the unreduced subsequence, as noted in Fig. 3.5 by the arrows. One advantage of that approach is an extreme reduction of computing time. On the downside this approach eliminates possibly important information concerning the duration of a sequence. To solve this issue the *logarithmic symbolic reduction* is now introduced as an extension of standard symbolic reduction. The idea is to compress a subsequence of equal consecutive symbols depending on its length, which is shown in the following equation using the logarithm naturalis:

$$|subsequence_{reduced}| = \lceil \ln(|subsequence|) \rceil \qquad (3.4)$$

In this case a *subsequence* contains only one kind of symbol and $|subsequence|$ states the length of that sequence. Here the basis of Euler's number was chosen, but it is possible to use any number > 1, which is why it is named $logBasis$ for further explanations. One goal of the logarithmic reduction is the improvement of similarity concerning the length information especially for dynamic situations. Nevertheless, the reduction in length (compression) is desired to be as high as possible, while preserving the relation between sequence and time. In the following paragraph the logarithmic symbolic reduction algorithm and its basic rules are described. For clarification new symbols X^i_j are introduced with i being the magnitude of the number of represented symbols X and j the index of symbol X in $subsequence_{reduced}$. Furthermore, $|X^i_j|$ denotes the number of represented symbols of type X by X^i_j in $subsequence$. Two rules are defined:

for $subsequence_{reduced} = X_0^0.X_1^1.\ldots.X_{n-1}^{n-1}.X_n^n.X_{n+1}^{n-1}.\ldots.X_{2n-2}^1.X_{2n-1}^0$

1: $|X_j^i| \le log\,Basis^{i+1}$

2: if X^n occurs, then $X^{n-1}, X^{n-2}, \ldots, X^0$ must also occur in $subsequence_{reduced}$

$$(3.5)$$

Algorithm 1 Logarithmic Reduction for compression of discretized time series.

1: **procedure** LOGARITHMICREDUCTION($subsequence, log\,Basis$)
2:     $newSymbols = nan(length(subsequence), 1)$
3:     $i = 0$
4:     $ruleViolation = true$
5:     $lasTD = 1$
6:     $lasBU = length(subsequence)$
7:     **while** $ruleViolation$ **do**
8:         $indexi = lasTD + floor(log\,Basis^i) - 1$
9:         $newSymbols(indexi) = i$
10:        $lasTD = indexi$
11:        $ruleViolation = checkRuleViolation(newSymbols, log\,Basis)$
12:        **if** $ruleViolation$ **then**
13:            $indexi = lasBU - floor(log\,Basis^{i+1}) + 1$
14:            $newSymbols(indexi) = i$
15:            $lasBU = indexi$
16:            $ruleViolation = checkRuleViolation(newSymbols, log\,Basis)$
17:        **end if**
18:        $i = i + 1$
19:    **end while**
20: **end procedure**

Table 3.2 Example for the logarithmic symbolic reduction to the basis of Euler's number using the *first* subsequence from Fig. 3.5

original	A	A	A	A	B	B	B	B	B	B	B	B	A	A	A	A
reduced	A_1^0		A_2^0		B_1^0		B_2^1				B_3^0		A_1^0		A_2^0	
represented symbols $|X|$	2		2		2		4				2		2		2	
$< e^x$	e^1		e^1		e^1		e^2				e^1		e^1		e^1	

Algorithm 1 follows these rules. Having a subsequence of equal consecutive symbols and a given logarithmic basis, the algorithm runs a WHILE loop. Within every iteration it at first inserts a new symbol at the front of the reduced subsequence.

If the rules, especially rule 1, are violated, the algorithm inserts another symbol at the end of the reduced subsequence and checks the violation again. With every new iteration the value of i increases by one and thus the inserted symbols X^i are able to represent more symbols X. The higher i is, the bigger is the gap between the inserted symbols, which is also depending on the variable $logBasis$. If there is no rule violation, the WHILE loop stops and the function outputs the reduced subsequence, which is named $newSymbols$ here. This variable has the same length as the input subsequence, but contains NaNs (Not a Number) at the indices of deleted symbols. Later the NaNs are eliminated and the corresponding indices are saved. The result is a high resolution (low compression) near to SVCs and a low resolution (high compression) in the center of long sequences of equal symbols. Note that the longer the subsequence, the higher the compression is.

In Table 3.2 an exemplary logarithmic reduction is performed based on the first sequence of Fig. 3.5 and using Euler's number as a basis. In the first row the original symbolic sequence is shown, while the second row displays the reduced sequence, received by applying logarithmic reduction. Each empty cell points to a symbolic value and its index, that is deleted by the algorithm in order to compress the sequence. For better comprehension rows three and four show the number of original symbols, that a new symbol represents, and the logarithmic value, which was relevant for the algorithm to place a new symbol and complement it with an exponent. The logarithmic reduction of sequences two and three are shown within the appendix (accessible in the Electronic Supplementary Material) in Table A.1 and Table A.2. The three sequences are reduced to:

1. $A^0 A^0 B^0 B^1 B^0 A^0 A^0$
2. $A^0 A^1 A^0 B^0 B^1 B^1 B^0 A^0 A^1 A^0$
3. $A^0 A^1 A^0 B^0 B^1 B^1 B^0 A^0 A^1 A^0$

Even though they have basically the same sequence of symbols ABA, they partly show different logarithmically reduced sequences. The first original sequence is much shorter than sequences two and three, which results in a reduced symbolic sequence varying from those of the other sequences. Consequently, sequences two and three can be detected as a pattern, due to their identical reduced sequence. Sequence one will not be assigned to the same pattern. By applying the logarithmic reduction algorithm it is now possible to differentiate between short- and long-term symbolic value changes, while maintaining a high compression. Furthermore, the sensitivity of this differentiation can be parametrized by varying the logarithmic basis.

Note that there are two possibilities to handle the new symbols X_j^i. At first every symbol X_j^i from the reduced sequence could be converted back to the original symbol X, to stay in the same symbolic space as before. But not converting them is also an interesting possibility. Even though the number of symbols D would multiply, it prevents the pattern discovery to compare symbols of different magnitudes of length. Because of this advantage, in the following evaluation of Chapter 5 the symbols are not converted back. After reducing the sequence, finally the pattern discovery can be applied .

3.4   Creating a Dictionary

This section was taken from previous work [70, 72] to a large extend. In the previous section techniques were described to transform real valued time series into symbolic time series. This symbolic time series can now be used to discover recurring patterns and create a dictionary. The dictionary is required to store the symbolic sequence of the discovered patterns as well as the original start and end index of every pattern and each member. Algorithms that are suitable to run on discretized time series were already enumerated and briefly explained in Section 2.1.3, which includes the text mining algorithm Sequitur. The following sections deal with two alternative algorithms, namely an adapted version of the Sequitur algorithm and a pattern enumeration algorithm. Both are able to fulfill the required task by identifying identical recurring symbolic sequences with variable length.

3.4.1   Adapted Sequitur Compression Algorithm

As briefly described in Section 2.1.3, Sequitur is a text compression algorithm that extracts a hierarchical structure from the symbolic sequence. In general it is an online algorithm able to perform pattern discovery in streaming applications. In Table 3.3 an exemplary execution of the algorithm is shown based on a symbolic sequence, that grows in every iteration by a new incoming symbol. In every iteration the extracted grammar is updated under consideration of two basic constraints: *digram uniqueness* and *rule utility*. The digram uniqueness says, that every digram (a pair of two adjacent symbols) is allowed to occur only once in the symbolic sequence. If a digram appears more than once, a new rule has to be created in the grammar. The new rule is given a new symbol, which replaces all occurrences of the digram in the symbolic sequence. The rule utility constraint says, that a rule has to occur at least twice in the grammar. Hence, if a new rule is created, the lower level rules need to be rechecked and eventually be deleted. This process can be observed in iteration 10 in Table 3.3.

Table 3.3 Execution of the original Sequitur algorithm on a sequence of symbols (illustration based on [69])

Symbol number	The string so far	Resulting grammar	Remarks
1	a	S → a	
2	ab	S → ab	
3	abc	S → abc	
4	abcd	S → abcd	
5	abcdb	S → abcdb	until now no digram found
6	abcdbc	S → abcdbc	*bc* appears twice
		S → aAdA A → bc	enforce digram uniqueness by replacing *bc* with *A*
7	abcdbca	S → aAdAa A → bc	
8	abcdbcab	S → aAdAab A → bc	
9	abcdbcabc	S → aAdAabc A → bc	*bc* appears twice
		S → aAdAaA A → bc	enforce digram uniqueness aA appears twice
		S → BdAB A → bc B → aA	enforce digram uniqueness by creating new rule for *aA*
10	abcdbcabcd	S → BdABd A → bc B → aA	*Bd* appears twice
		S → CAC A → bc B → aA C → Bd	enforce digram uniqueness *B* is only used once
		S → CAC A → bc C → aAd	enforce rule utility by deleting rule *B*

In this work an online application of pattern discovery is not necessary, which opens up the opportunity to implement an offline version of the Sequitur algorithm that executes even faster. The pseudocode is shown in Algorithm 2. As it is an offline algorithm, the input is the whole symbolic time series T_{disc}. The core of the algorithm is implemented in lines 4 to 6. These three lines calculate a statistic of all the digrams included in the symbolic time series, their absolute frequency and location. In line 4 the symbolic time series is transformed to a matrix of the size $(length(T_{disc}) - 1) \times 2$. The function $formMatrix$ reproduces the symbolic time series once and forms a matrix by shifting the reproduced sequence by one step. This leads to a matrix with rows corresponding to consecutive symbols of the symbolic time series. In line 5 the matrix is sorted, while both columns of the same row stay connected. Hence, identical digrams are now grouped together in consecutive rows within the matrix. Based on this matrix the function $getSymbolicRepr$ can easily create the required statistics.

At this point of the execution there is not only the information that a digram is occurring twice, like in online Sequitur, but there is knowledge about all digrams and their absolute frequencies. Hence, there is the possibility to choose which digram to replace next based on different objectives. In this implementation the next digram is the one with the highest absolute frequency. But it would also be possible to choose e.g. the one that covers the biggest part of the original time series. In lines 8 to 10 the best digram is chosen, added to the dictionary and replaced at all locations within the symbolic time series. This procedure is continued iteratively until no digram is found with a minimum count of 2, while the minimum count parameter $MinCount$ can also be parametrized. This marks the non-violation of digram uniqueness constraint.

Algorithm 2 Adapted Sequitur for identical sequence discovery.

1: **procedure** ADAPTEDSEQUITUR(T_{disc}, $MinCount$)
2:    $empt = false$
3:    **while** ¬$empt$ **do**
4:       $tsMatrix = formMatrix(T_{disc}, 2)$
5:       $[tsMatrixSorted, index] = sort(tsMatrix)$
6:       $[symbolicRepr, num, loc] = getSymbolicRepr(tsMatrixSorted)$
7:       **if** $max(num) > MinCount - 1$ **then**
8:          $[v, i] = max(num)$
9:          $add2Dictionary(symbolicRepr(i), loc(i))$
10:          $T_{disc} = replaceByNewRule(T_{disc}, loc)$
11:       **else**
12:          $empt = true$
13:       **end if**
14:    **end while**
15: **end procedure**

3.4.2 Pattern Enumeration Algorithm

The pattern enumeration algorithm solves the same task as Sequitur. It repeatedly scans the whole symbolic time series with different symbolic lengths, but without compressing it. The approach is still similar to the Sequitur algorithm, as shown in Algorithm 3. In lines 6 to 8 the same statistics are calculated, with only one minor difference. With every iteration of the WHILE loop the symbolic length to be searched $lenPattern$ increases by one, while in every iteration of Sequitur only digrams are searched. Hence, the matrix formed in line 6 has the size of $(length(T_{disc}) - lenPattern + 1) \times lenPattern$ containing $lenPattern$ consecutive symbols in one row. In lines 9 to 13 each of those symbolic sequences occurring at least twice, or in general $MinCount$ times, is added to the dictionary. If in an iteration of the WHILE loop no symbolic sequence with that property is found, the algorithm stops. It is also possible to limit the minimum and maximum symbolic length ($MinPatternSize$, $MaxPatternSize$) to be searched.

In general this approach is more naive, because it just enumerates all the included patterns. There are basically three reasons, why it still makes sense to implement pattern enumeration instead of Sequitur. First of all, in the process of Sequitur rules can be covered by other rules. In Table 3.4 such an example is shown by compressing the same symbolic sequence twice, but slightly varying the order of digrams to be replaced by new rules. Obviously, there is the recurring pattern abc that should be discovered by the algorithm. Within the first order of replacement the algorithm starts by discovering rule $A \rightarrow ab$, followed by $B \rightarrow Ac$. By unfolding rule B, we see that the algorithm was able to find the desired pattern abc. But if the order of replacements is started by surrogating rule $C \rightarrow cb$, the algorithm ends up not discovering the pattern abc, because is was covered by rule C. This leads to the second reason to choose pattern enumeration over Sequitur, which is even more general. In Sequitur the pattern discovery result depends on the order of replacements. The pattern enumeration algorithm executes without a sequence dependency and prevents the pattern discovery from covering rules.

The third reason to choose the pattern enumeration over Sequitur is the advantage of variability in postprocessing. In the postprocessing step relevant patterns are chosen from the dictionary. The definition of pattern relevance depends on the users intention. To give the user the greatest possible freedom in choosing relevant patterns, it is beneficial to use pattern enumeration. It outputs a complete list of patterns, while Sequitur itself already makes a decision to limit the number of patterns in the dictionary by its replacing procedure.

3.5   Postprocessing—Selection of Relevant Patterns

The selection of relevant patterns after their discovery is an important postprocessing step. Without postprocessing the result of pattern discovery would just be a huge confusing amount of patterns. Especially in case of the pattern enumeration algorithm, the result is just a list of all patterns included in the data set. Even though the Sequitur algorithm has a built-in selection mechanism, in most cases postprocessing is still necessary. Depending on the application and the users intention only a subset of patterns may be classified as relevant, while the definition of relevance can vary a lot. A common technique to handle large amounts of patterns is to define a ranking function based on one or multiple of the following figures:

- Length of patterns (symbolic or original)
- Absolute frequency
- Coverage of (symbolic or original) time series
- Dynamics (e.g. number of different symbols in a pattern)
- Periodicity of pattern occurrences

Table 3.4 Weakness of Sequitur. The results depend on the sequence of replacements. Rules can be covered by other rules

Order of replacement	The string so far	Resulting grammar	Remarks
1st	…abcbabc …	S → …abcbabc …	*ab* appears twice
		S → …AcbAc … A → ab	Enforce digram uniqueness by replacing *ab* with *A*. *Ac* appears twice
		S → …BbB … A → ab B → Ac	enforce digram uniqueness by replacing *Ac* with *B*
2nd	…abcbabc …	S → …abcbabc …	imagine *cb* appears twice
		S → …abCabc … C → cb	Enforce digram uniqueness by replacing *cb* with *C*. *ab* appears twice
		S → …ACAc … C → cb A → ab	enforce digram uniqueness by replacing *ab* with *A*

Algorithm 3 Pattern Enumeration for identical sequence discovery.

1: **procedure** PATTERNENUM(T_{disc}, $MinCount$, $MinPatternSize$, $MaxPatternSize$)
2:    $lenPattern = MinPatternSize - 1$
3:    $empt = false$
4:    **while** $\neg empt$ **do**
5:       $lenPattern = lenPattern + 1$
6:       $tsMatrix = formMatrix(T_{disc}, lenPattern)$
7:       $[tsMatrixSorted, index] = sort(tsMatrix)$
8:       $[symbolicRepr, num, loc] = getSymbolicRepr(tsMatrixSorted)$
9:       **for** $p \leftarrow 1, length(symbolicRepr)$ **do**
10:          **if** $num(p) > MinCount - 1$ **then**
11:             $add2Dictionary(symbolicRepr(p), loc(p))$
12:          **end if**
13:       **end for**
14:       **if** $isempty(find(num > 1)$ **or** $lenPattern == MaxPatternSize$ **then**
15:          $empt = true$
16:       **end if**
17:    **end while**
18: **end procedure**

In contrast to the ranking approach, this section focuses on selecting the most representative patterns with respect to the RDC analysis. More formally, the task is to find the smallest set of patterns, that in total covers the whole (or at least the majority) of the original data set. An approach to solve this *set cover problem* is proposed in Section 3.5.2. Before diving into this problem, in Section 3.5.1 an additional filtering method is introduced, that is necessary because of the additional preprocessing techniques proposed in Section 3.3.

3.5.1   Filtering of Irrelevant Patterns

Due to the usage of logarithmic reduction within the preprocessing it makes sense to include an additional filtering method before finally selecting relevant patterns in Section 3.5.2. As already described in Section 3.3.3 the implementation of new symbols X^i instead of just X is just an algorithmic technique to enable the differentiation between patterns with varying durations. Hence, these auxiliary symbols have value for the pattern discovery algorithm, but not for the user itself directly. Imagine the users running the pattern discovery at first with standard reduction. They choose a $MinPatternSize = 3$, because they want the patterns to contain a minimum amount of dynamic behavior. The algorithm finds e.g. symbolic sequences like ABC or ABA. Afterwords they decide to try again with logarithmic reduction to explicitly take into account the duration of patterns. After the preprocessing there may be a symbolic sequence that partly looks like that: $A^0.A^1.A^2.A^1.A^0$. Every symbol in this sequence is classified as A, but with different magnitude of represented original samples. Hence, all the related original values in this small subsequence range within the discretization boundaries of A. With a $MinPatternSize$ of three, patterns that can be found within this subsequence are the following: $\{A^0.A^1.A^2\}$, $\{A^1.A^2.A^1\}$, $\{A^2.A^1.A^0\}$. All these patterns are only describing constant behavior of the original signal(s). This behavior may cause confusion, because the users still expected patterns of at least three real symbolic value changes.

To gain stability and transparency of the pattern discovery when varying parameters, an additional filtering of these irrelevant patterns is proposed. Every entry in the dictionary is checked again regarding the following property: Are there at least $MinPatternSize - 1$ true value changes within a pattern? A false value change is a change in magnitude, but not in the underlying symbol itself. Every pattern containing the minimum amount of true value changes is kept in the dictionary, every other is deleted.

3.5.2 Selection of Representative Patterns

Some parts of this section were taken from previous work [35]. The task of finding the most relevant patterns, or a representative set of patterns, can be classified as *set cover problem*:

Definition 8 *Given a set U and n subsets S_j of U, the set cover problem formulates the task of finding the smallest number $k \leq n$ of subsets to cover all the objects in U by the union of all k subsets.*

The problem was proven to be NP-complete by Karp in 1972 [47]. Translated to selecting a representative set of patterns, U contains all the samples of the original time series, that is input to pattern discovery. A pattern with every of its members can be described by every sample that is part of the pattern. Hence, a pattern j is a subset S_j. Every sample in the described scenario has a unique index, defined by the sequence in U, i.e. the original time series. Therefore U as well as every subset S_j is a vector of indices referencing to a sample in the time series.

Due to the high complexity of the set cover problem, a greedy algorithm is now proposed to find the smallest set of patterns that matches the definition. In the following at first the classic greedy approach is described, followed by some simplification techniques, the implementation of the greedy algorithm as well as different extensions to weighted greedy approaches.

Greedy Set Cover Algorithm
The first application of a greedy strategy to solve the set cover problem was introduced in 1979 by Chvatal in [13]. The greedy strategy, in general, only provides an approximate solution, but is still the dominating approach to solve the problem. In [31] Grossman and Wool performed a study of nine different algorithms, including different types of greedy heuristics and one NN-based approach, to solve the set cover problem with multiple randomized and combinatorial test data sets. They compared these algorithms regarding the distance between their discovered solutions and the optimal solution as well as regarding the time complexity. One of the authors conclusions was, that the best choice of algorithm depends on the type of data set. In contrast to most of the algorithms, the classic Greedy algorithm provided good results in both types of data sets, being within the top 3 in both cases, while at the same time being one of the fastest algorithms. Due to the robustness and fast execution, we choose the classical greedy approach for the selection of patterns.

The greedy set cover heuristic is shown in Fig. 3.6. In Fig. 3.6 a) the initial situation is shown. The black rectangle represents all the indices of the time series that are supposed to be represented by patterns. Additionally, four exemplary patterns and their index sets are shown as four bubbles with different hatching. The greedy algorithm chooses one pattern in every iteration to be in the representative pattern set on the basis of their count of not yet represented indices. Hence, in a) pattern 2 is chosen, because no index is already represented and its bubble is the biggest one. In the second iteration, shown in b), the filled area, which used to be the bubble of pattern 2, is already represented and does not need to be represented again. This is why the area of pattern 4 reduces by the intersection of pattern 2 and 4. Still the area of pattern 4 is chosen, because of its left-over size. In c) the represented area increases by the area of pattern 4 and now shows the union of pattern 2 and 4. This time the areas of pattern 1 and 3 reduce by their intersection of pattern 4. Continuing in this manner a representative pattern set is chosen as an approximate of the set cover problem.

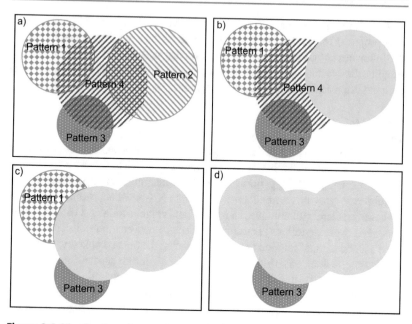

Figure 3.6 Visualization of pattern index sets and the execution of a greedy algorithm to select relevant patterns. a) initial state; b) - d) three iterations of greedy algorithm [5]

Simplifications

As the problem is NP-complete and the algorithm is required to be applied to very large data sets, techniques are necessary to simplify the task. The first simplification adds a degree of freedom by introducing the threshold $\alpha \in [0, 1]$. It defines a break condition for the greedy algorithm and states a minimum relative coverage to be reached when running the greedy algorithm. Hence, an α-value of 0.9 requires at least 90% of the time series to be represented by patterns. Setting α to a value < 1 results in an early termination of the algorithm and therefore saves computing time. This threshold is additionally justified by the fact that the pattern discovery is required to detect recurring patterns. Due to the complexity of real-life applications, it is likely that data sets contain a small number of subsequences that are unique. Hence, it may be impossible to represent the time series by only recurring patterns.

[5] „Journal of Energy Storage, vol. 38, Heinrich et al., Unsupervised data-preprocessing for Long Short-Term Memory based battery model under electric vehicle operation, pp. 102598, (2021)", with permission from Elsevier Science & Technology Journals. The publication is available at ScienceDirect through https://doi.org/10.1016/j.est.2021.102598.

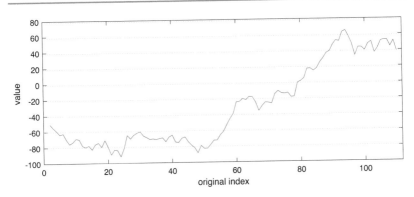

Figure 3.7 Exemplary part of a time series including discretization boundaries

Table 3.5 Exemplary part of a symbolic time series based on Fig. 3.7 including the corresponding original indices and the number of represented samples

Index of symbolic TS	1	2	3	4	5	6	7	8	9	10
Corresponding index of original TS	1	4	20	25	56	58	71	78	85	89
Symbolic TS	3	2	1	2	3	4	5	6	7	8
Number of represented samples	3	16	5	31	2	13	7	7	4	20

Another technique deals with an optimization that results from the compressed time series. After preprocessing in Section 3.3 the symbolic time series is much shorter than the original time series. In order to save computing time when applying set operations, it is beneficial to use the shorter symbolic time series and its indices. The results can easily be translated back into the original space, because every symbol in the symbolic time series represents a known number of original samples. The needed information is stored in the process of symbolic reduction described in Section 3.3.3. An exemplary output of the preprocessing is shown in Table 3.5, which corresponds to the univariate time series in Fig. 3.7. In this case the preprocessing included the use of discretization, hysteresis and standard symbolic reduction. The resulting symbolic time series is shown in row 3. Every symbol is assigned to a continuously increasing index, a corresponding index of the original time series, as well as a number of represented samples.

Implementation of Greedy Algorithm

The implementation of the greedy set cover algorithm including the simplifications of the previous section is shown as pseudocode in Algorithm 4. The function requires a dictionary *Dict*, a sequence matrix *Seq* as shown in Table 3.5, the threshold α and an optional specification of the *weightMode*, which will be described in more detail in the next section. The default mode activates the classic greedy algorithm as described previously. In line 2 three types of index vectors U, S_j and R_j for every pattern j are created equivalent to the previous description. R_j additionally stores indices that are already represented separately for every pattern j. At this point of the algorithm every entry of R is empty. In line 3 all figures are calculated, that are necessary for the decision of the next relevant pattern. In case of the classic greedy algorithm, only one figure is used, describing the coverage of a pattern of not yet represented indices:

$$coverage_j = |S_j \cap R_j| \tag{3.6}$$

Because the algorithm is based on the compressed time series, $coverage_j$ has to be translated into the original space. This is also part of the function *calcFigures*. After some initial steps in line 4 and 5 the WHILE loop runs until the overall coverage of the original time series exceeds the threshold α. Based on the chosen figures, at the beginning of each iteration (line 5) the next patterns are selected. Note that in this step, in contrast to the classic greedy algorithm, it is possible to select multiple patterns in one iteration. In line 8 these patterns are added to a new reduced dictionary *redDict*. In the tenth line U is adapted by the index sets of the currently selected patterns. After that the *overallCoverage* is updated and all selected patterns are deleted from all variables, except for the reduced dictionary obviously. The function *calcOverlap2TS* finally updates the index vectors S and R as well as the figure describing the coverage for the iteration.

Because the majority of computing time results from the calculation of coverage and the update of index vectors, it is useful to take a closer look into the function *calcOverlap2TS*. Based on S, $Snext$, R and Seq it runs through a FOR-loop to update the vectors and figures as shown in Algorithm 5. First of all in line 3 the function *isMember* checks for every index in S_j if it is also a member of the previously selected patterns by intersection of both sets. Based on this overlap both index vectors S_j and R_j are updated. S_j is reduced by every index that is already represented. R_j is extended by the previously calculated overlap. By applying the function *calcRelativeDescription* the $coverage_j$ is translated to the original space in line 6:

$$overlap2Original_j = \frac{\sum Seq(4, S_j)}{|U|} \tag{3.7}$$

In the same manner the figure $overlap2Represented_j$ can be calculated. It additionally describes the relative coverage of already represented parts of the time series for each pattern. When applying the classic greedy approach this figure is not used. Nevertheless, it contains valuable information, which is why it is considered for the weighted greedy algorithm. It is formally defined as:

$$overlap2Represented_j = \frac{\sum Seq(4, R_j)}{\sum Seq(4, \bigcup_{\forall j} R_j))} \tag{3.8}$$

Algorithm 4 Greedy algorithm for the selection of relevant patterns.

1: **procedure** SELECTRELEVANTPATTERNS($Dict, Seq, \alpha, weightMode$)
2:    $[S, R, U] = createIndexVectors(Dict, Seq)$
3:    $fig = calcFigures(Dict, S)$
4:    $overallCoverage = 0$
5:    $initredDict$
6:    **while** $overallCoverage < \alpha$ **do**
7:       $nextPatterns = selectNext(fig, weightMode)$
8:       $redDict = add2Dictionary(nextPatterns, Dict, redDict)$
9:       $Snext = S(nextPatterns)$
10:      $U = setdiff(U, Snext)$
11:      $overallCoverage = 1 - calcRelativeDescription(U, Seq)$
12:      delete already chosen patterns from all variables
13:      $[S, R, fig] = calcOverlap2TS(S, Seq, Snext, R)$
14:   **end while**
15: **return** $redDict$
16: **end procedure**

Weighted Greedy Algorithm
As already mentioned, it may be useful to include more information in the selection process, which is already prepared in Algorithm 4. In line 3 multiple figures for all patterns are calculated. Their usage is defined by the variable $weightMode$. Beside the previously defined figures $overlap2Original$ and $overlap2Represented$ it is possible to use the following ones:

- Absolute number of members of each pattern ($numMembers$): This option can be used to set more focus to short high frequent patterns instead of rare long patterns.

Algorithm 5 Update index vectors within greedy algorithm by chosen patterns and output of new overlaps.

1: **procedure** CALCOVERLAP2TS(S, Seq, $Snext$, R, fig)
2:    **for** $j = 1 : length(R)$ **do**
3:       $[isOverlap] = isMember(S(j), Snext)$
4:       $R(j) = sort([R(j); S(j)(isOverlap)])$
5:       $S(j) = S(j)(\neg isOverlap)$
6:       $overlap2Original(j) = calcRelativeDescription(S(j), Seq)$
7:       $overlap2Represented(j) = calcRelativeDescription(R(j), Seq)$
8:    **end for**
9:    update overlap2Original and overlap2Represented in fig
10: **return** S, R, fig
11: **end procedure**

- Symbolic length of a pattern ($symLength$): In contrast to the previous one, this figure can be used to favor symbolically long patterns.
- Number of different symbols in a pattern ($uniqueSymLength$): The usage of this figure sets more focus to dynamic patterns.

While the figures $overlap2Original$ and $overlap2Represented$ need to be updated every iteration, the figures $numMembers$, $symLength$ and $uniqueSym$ $Length$ are constant. In general, all these figures can be combined in any way dependent on the application and the users intention. In Fig. 3.8 four frequently used possibilities are shown for the selection process with different weighting functions. Every point represents one pattern and its calculated figures in one iteration of the greedy algorithm. In Fig. 3.8 $overlap2Original$ and $overlap2Represented$ are used as a basis for the selection process. For every pattern a weight is calculated by the following formula:

$$weight_j = overlap2Original_j \times (1 - overlap2Represented_j) \qquad (3.9)$$

As indicated by the isolines, preferred patterns have a high overlap to not yet represented indices and a low overlap regarding already represented indices. The isolines show possible values for the weight of patterns. In Fig. 3.8b another two-dimensional selection is shown. The $overlap2Original$ is complemented by $uniqueSymLength$ and a weight is calculated by:

$$weight_j = overlap2Original_j \times \frac{uniqueSymLength_j}{max(uniqueSymLength)} \qquad (3.10)$$

The variable *uniqueSymLength* is normalized by its maximum in order to serve comparable weights between 0 and 1. Additionally a combination of three figures is used in Fig. 3.8d with the weighting function:

$$weight_j = overlap2Original_j \times \frac{uniqueSymLength_j}{max(uniqueSymLength)}$$

$$\times (1 - overlap2Represented_j) \tag{3.11}$$

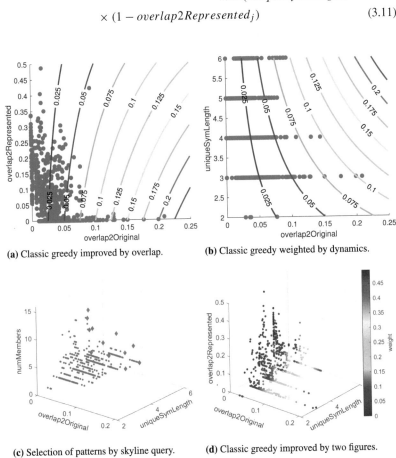

(a) Classic greedy improved by overlap.

(b) Classic greedy weighted by dynamics.

(c) Selection of patterns by skyline query.

(d) Classic greedy improved by two figures.

Figure 3.8 Exemplary weighting for selection of next patterns in one iteration of the greedy algorithm

These three weighting functions have in common, that the values of different figures are projected onto one dimension. In every iteration the pattern with the highest weight is chosen to be part of the representative pattern set. While in theory a lot of different figures and weighting functions are conceivable, it is mandatory to include *overlap2Original* due to the break condition in Algorithm 4.

As already mentioned, it is possible to select multiple patterns in one iteration. In Fig. 3.8c such an application is shown. In contrast to the previous examples, in this case no weighting functions needs be defined. The selection is done by utilizing a *Skyline-Query*. The skyline operator is a multi-criteria optimization based on dominance relations between tuples. In this case a tuple is a pattern defined by the values of the chosen figures. A tuple, or a pattern, is part of the skyline if it is not dominated in all dimensions [93]. The idea is to select all the patterns that are part of the skyline. In Fig. 3.8c all patterns marked as diamond are part of the skyline based on three criteria *overlap2Original*, *numMembers* and *uniqueSymLength*. This approach has the advantage of computing time optimization due to the lowered number of iterations that need to be done. On the downside the amount of indices that are covered multiple times increase, which should be avoided in case of RDC analysis. On the other hand it is perfectly suited to get a first glance of a data set, because it outputs a little bit of everything.

Table 3.6 Variants of weighted greedy to be used in the evaluation

Names	Function
standard	$overlap2Original_j \times (1 - overlap2Represented_j)$
static	$overlap2Original_j \times (1 - overlap2Represented_j) \times \dfrac{symLength_j}{max(symLength)}$
dynamic	$overlap2Original_j \times (1 - overlap2Represented_j) \times \dfrac{uniqueSymLength_j}{max(uniqueSymLength)}$
skyline	skyline index of overlap2Original, overlap2Represented, uniqueSymLength and numMembers

In Section 5.1 different possibilities of the weighted greedy approach will be evaluated against each other as well as against results without postprocessing. The variants that will be used in the evaluation are named as *standard*, *static*, *dynamic* and *skyline* weighted greedy. Their functions are shown in table Table 3.6.

Pattern-based Representative Cycles

4

The pattern discovery is able to solve a variety of problems related to time series as it is able to adapt to the use cases requirements. To exemplarily demonstrate its benefit, this chapter deals with the idea and implementation of pattern-based RDCs as already outlined in Section 2.3. At first, to recapture the actual problem that an RDC construction method has to solve, the initial situation is described: Large amounts of time series data is recorded showing the usage of a vehicle or even a fleet of vehicles over a long period of time. Due to the size of this data set it is difficult to mine information regarding their usage. An RDC construction method aims to identify or create a short RDC that is able to represent the original data. Furthermore, the task is to construct the shortest possible RDC while maintaining the highest amount of representativeness. In doing so, for example in the vehicle engineering process it is possible to either reduce the testing effort or add representative data to enhance it. The application of RDCs was already described in more detail in Section 2.2.1.

The idea of pattern based RDCs is to use pattern discovery to identify all patterns, that in total represent the majority of the data. Hence, the representativeness is defined by discovered patterns and their frequency. This concept is described in more detail in the following Section 4.1. Based on the definition of representativeness two approaches are proposed. In Section 4.2 the identification of RDCs based on representative patterns is described, which is classified as a full trip method. Beyond that, in Section 4.3 an approach is proposed to semi-synthetically construct RDCs by concatenating representative patterns.

F. K. D. Noering, *Unsupervised Pattern Discovery in Automotive Time Series*, AutoUni – Schriftenreihe 159, https://doi.org/10.1007/978-3-658-36336-9_4

4.1   Pattern-based Statistics

Before going into detail regarding the identification and construction of RDCs, it is necessary to describe the pattern-based statistics as its theoretical foundation. While the representativeness as well as the characteristic parameters in literature are mostly based on expert knowledge, by the use of pattern discovery it is now possible to establish an unsupervised approach to define the representativeness. In Section 4.1.1, at first, the characteristic parameters are introduced, which describe the link between pattern discovery and representativeness. Based on these parameters, in Section 4.1.2 four basic use cases for RDCs are differentiated. Moreover, in Section 4.1.3 a pattern-based distance measure is introduced, which is able to express the quality of an RDC. As the pattern-based RDC construction and identification is made to be a generalized approach, the users have to be able to input their requirements. In Section 4.1.4 the influence of pattern discovery and the possibility to customize the method is discussed.

4.1.1   Characteristic Parameters

Characteristic parameters (CP) are features that define the representativeness. Usually, they are based on expert knowledge and are customized for each use case. Hence, when working on an RDC for another application it is necessary to develop a function that captures the representativeness. Due to the overall approach of this work to develop an RDC construction method that is highly unsupervised, it is evident to also introduce unsupervised CPs. This section proposes CPs that are independent from expert knowledge and, instead, utilize the unsupervised discovered patterns with absolute and relative frequency.

As the pattern discovery is applied to the data set, it discovers recurring patterns. By means of the greedy postprocessing technique, the method selects a set of patterns, that in total covers the majority of the time series. This is parametrized by setting α, which describes the proportion of the time series that at least needs to be covered by patterns. In this application of RDC construction it is necessary to set high values of α, usually 0.9 or higher. In general, the representativeness of the RDC will increase when increasing α.

A time series data set can be described by its patterns, which is visualized in Figure. 4.1. In Fig. 4.1a it shows a bar chart of 21 discovered patterns by their absolute frequency (filled bars) as well as their relative coverage (framed bars). The symbolic sequences of these patterns are written on the x-axis. These patterns were discovered in an exemplary time series with $\alpha = 0.9$. A more detailed definition of the CPs *absolute pattern frequency* and *relative pattern coverage* follows:

Definition 9 APF_j, the absolute pattern frequency of pattern j, expresses the absolute number of members of pattern j found in the time series. Accordingly, the absolute pattern frequency spectrum (APF-spectrum) includes the APF_j of every pattern j.

Definition 10 RPC_j, the relative pattern coverage of pattern j, expresses the number of indices covered by the pattern including all its members in relation to the length of the time series. In this case it does not matter if a portion of those indices is also covered by other patterns. Accordingly, the relative pattern coverage spectrum (RPC-spectrum) includes the RPC_j of every pattern j found in the time series.

Hence, a data set can be characterized by the APF- and RPC-spectrum. Note that there might be an overlap between the chosen patterns, which leads to a cumulative RPC-spectrum that rises above 1 (see Fig. 4.1b).

4.1.2 Use Case Differentiation

One important question for RDC construction is: What is the desired APF- and RPC-spectrum for the RDC? As already described, one objective of RDCs is data reduction. Hence, the APF-spectrum of the RDC needs to be on a lower level than the spectrum of the input data. In general, there are four different possibilities to achieve this goal, that basically relate to four different use cases.

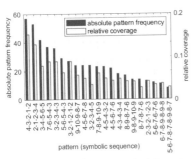
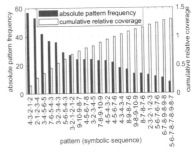

(a) Absolute pattern frequency and relative pattern coverage of every pattern.

(b) Absolute pattern frequency and cumulative relative pattern coverage.

Figure 4.1 Description of an exemplary time series by its patterns and their frequency

Use Case 1—Evenly Distributed Frequency

First of all, it is possible to reduce the APFs of every pattern to a small fixed number. This may be required in engineering applications with the goal to test a device with every possible situation that occurs in real life, no matter how frequently these situations are occurring. Hence, the desired APF-spectrum can be reduced to a constant value, as visualized on the left hand side of Figure. 4.2. The APF-spectrum of the input time series is shown as filled bars, while the desired APF-spectrum of the RDC is marked as framed bars. According to the desired APF-spectrum a theoretical RPC-spectrum can be calculated:

$$RPC_j = \frac{APF_j \times meanLength_j}{\sum_j (APF_j \times meanLength_j)} \qquad (4.1)$$

The variable $meanLength_j$ contains the average length of every pattern j. The RPC-spectrum is just theoretical, because at this points it is only a set of patterns that are chosen to be in the RDC. It is not yet a connected driving cycle. The transformation from a set of patterns to an RDC will be described later in Section 4.3. Note that the RPC-spectrum of the RDC in this use case also tends to show an even distribution, as shown in Fig. 4.2b. Nevertheless, it differs from an even distribution due to the variable lengths of patterns. This technique has the advantage of a massive data reduction.

Use Case 2—Representative Frequency

In addition to the requirement of covering every possible pattern, it may be necessary in some applications to maintain the distribution of the RPC-spectrum. This can be achieved by setting the APF of the RDC according to the APF in the original data for each pattern. Due to the fact that every pattern has to be included in the RDC at least once and the overall goal is to create the shortest possible RDC, the APF in the RDC of the least frequent pattern in the original data can be set to one. Accordingly, the *norm factor* is set to the value of the least frequent pattern. Based on this norm factor, the whole APF-spectrum of the RDC can be calculated. The result is shown in Figure. 4.3a for the APF and in Fig. 4.3b for the RPC. Due to the compression, the relation between the APFs in the original data can usually not be perfectly maintained in the RDC. This causes the RPC-spectrum of the RDC to not perfectly fit with the RPC-spectrum of the original data. But nevertheless, it shows that this technique has the potential of approximately maintaining the RPC-spectrum and therefore the representativeness.

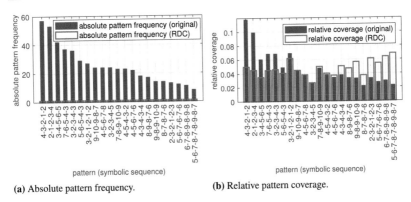

(a) Absolute pattern frequency.   (b) Relative pattern coverage.

Figure 4.2 Comparison of absolute pattern frequency and relative pattern coverage in original data as well as in the RDC. An evenly distributed pattern frequency is chosen for the RDC (use case 1)

Setting the least frequent pattern to one within the representative pattern set makes sense at the first view. But in practice this may result in a large representative set and subsequently an unnecessarily long RDC, especially in cases with rare patterns still covering a lot of the time series due to their duration. A common solution is to increase the norm factor. Unfortunately, this causes some patterns to be set to zero in the representative pattern set. Due to the importance of all patterns it is highly desirable to keep every pattern at least once in the RDC. This is why these patterns will automatically be set to one. In Section 5.2 this trade-off between representativeness and size of the RDC will be evaluated in more detail.

Use Case 3—Median Trip
Use case 3 has a slightly different basis in contrast to the previous use cases. While the previous use cases assumed one global APF- and RPC-spectrum for the original data, use case 3 is based on multiple spectra. Here the original data is not just one large time series, but many trips. This is because of the slightly different task of use case 3: Find the most representative trip in the set of all trips from the original data. This use case especially requires the RDC to be a real trip, that in fact was driven. In contrast to this, the previous use cases consider synthetic cycles that are designed to be representative.

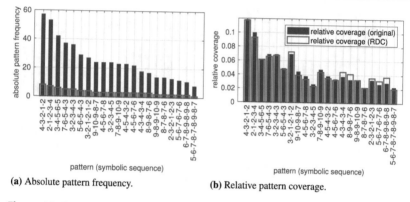

(a) Absolute pattern frequency.   **(b)** Relative pattern coverage.

Figure 4.3 Comparison of absolute pattern frequency and relative pattern coverage in original data as well as in the RDC. A representative pattern frequency is chosen for the RDC (use case 2)

To satisfy the requirements from use case 3, first the original data is cut into trips. This could, for instance, be done by extracting micro trips like already described in Section 2.2.3. Most of these described methods simply cut the velocity signal at zero velocity to extract micro trips. Later in this work another method to cut data into trips is proposed. The pattern discovery is usually applied to the uncut original data. The global set of patterns including their members then needs to be assigned to the trips. Hence, a set of patterns for every trip is extracted. This process will be explained in more detail in Section 4.2. These sets can then be transformed into APF- and RPC-spectra, that are visualized in Figure. 4.4. It shows 10 exemplary trips by their APF- (Fig. 4.4a) and RPC-spectra (Fig. 4.4b) as dotted lines. The trip that is closest to the center is chosen to be the RDC. In general, this can be done based on the APFs or the RPC. These median trips are highlighted in the figure as dashed lines.

As an alternative, the RDC can also be selected by comparing every trip with the overall RPC-spectrum of the original data. In this case the trip with the minimum distance regarding the original data would be chosen to be the RDC. In theory this is not equivalent to comparing the trips with their average value, due to the varying length of the trips. Nevertheless, the results should be similar in practice.

Use Case 4—Average Trip
In equivalence to the median trip of use case 3, it is also possible to derive an average APF-spectrum and then synthetically construct the RDC as in use cases 1 and 2.

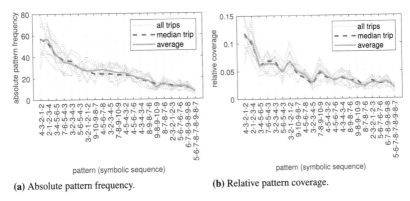

(a) Absolute pattern frequency.   (b) Relative pattern coverage.

Figure 4.4 Comparison of absolute pattern frequency and relative pattern coverage in original data as well as in the RDC. The original data is cut into trips and shown as dotted lines. The median (dashed line, use case 3) or average (solid line, use case 4) trip can be chosen to be the RDC

This average APF is marked as solid line in Fig. 4.4. In addition to use case 2, this RDC would be representative in its length regarding all trip lengths.

4.1.3   Pattern-based Distance Measure

In previous sections pattern-based CPs were introduced. Furthermore, varying requirements derived from different use cases regarding these statistics were described. It is now possible to introduce distance measures based on these statistics in order to compare the quality of different RDCs. The following distance measures can be used for the comparison of RDCs that aim for a good match in the RPC-spectrum. Hence, they are applicable for use cases 2 to 4.

In Figure. 4.5 the development of these distance measures is visualized based on the CPs. First of all, Fig. 4.5a shows the APF values of the original data and three RDCs. RDC 1 and 2 are created by the approach of use case 2, while RDC 3 is derived from the approach of use case 1. RDC 3 is used as a counterexample, i.e. a bad RDC, that does not match the RPC-spectrum of the original data, due to the fact that it is not designed to match. In Fig. 4.5b the corresponding RPC-spectra are shown, that are calculated according to Equation (4.1). Visually it is evident, that RDC 3 differs from the original data, especially in case of the first six and last six patterns. The distance measure should reflect the difference between RDC and

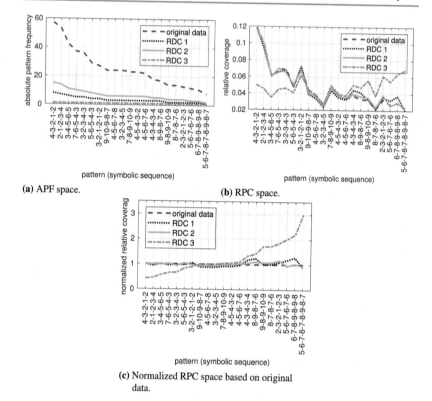

(a) APF space.

(b) RPC space.

(c) Normalized RPC space based on original
     data.

Figure 4.5 Comparison of three different RDCs regarding the original data in different spaces

original data within the RPC space. The better the RDC is, the smaller the distance
should be. In general, the L_k norm can be used for distance measurements with
regard to the RPC of the original data (RPC^{orig}) and averaged by the number of
patterns ($num\,Patterns$):

$$DM^k = \frac{\left(\sum_j \left| RPC_j - RPC_j^{orig} \right|^k \right)^{1/k}}{num\,Patterns} \tag{4.2}$$

The most popular L_k norm is the Euclidean distance with $k = 2$. It sums up the
squared distances of every pattern j and subsequently extracts the root:

$$DM^2 = DM^{Eucl} = \frac{\sqrt{\sum_j \left(RPC_j - RPC_j^{orig}\right)^2}}{numPatterns} \tag{4.3}$$

Unfortunately, every L_k norm has some issues with high-dimensional data, formally known as *curse of dimensionality*, which was studied in [2]. It says, that in high-dimensional data the distance metrics become meaningless. For example in nearest neighbor applications the relation between the nearest and the farthest neighbor to a given query converges to one with increasing dimensionality. This can be shown by means of the *relative contrast* with DM_{min}^k being the nearest and DM_{max}^k the farthest neighbor:

$$RelativeContrast = \frac{DM_{max}^k - DM_{min}^k}{DM_{min}^k} \tag{4.4}$$

The relative contrast defines the meaningfulness of a distance metric. The proposal of [2] was to reduce k when using the L_k norm in high-dimensional spaces. They proved that the meaningfulness 'worsens faster with increasing dimensionality for higher values of k'([2]). This is why the authors additionally introduced fractional distance metrics, that use values of k below one ($0 < k < 1$). In doing so, they were able to increase the meaningfulness of distance metrics in high-dimensional spaces. On the downside fractional distance metrics violate the triangle inequality, which is why they are no proper distance metrics. Still, this is not a problem in the use case of RDC construction, which is why the fractional distance metrics can be considered as useful.

Within the application of RDC construction, there is another problem with any L_k norm. This problem emerges with increasing values of α. As α increases, more patterns are discovered, which usually leads to more patterns that have low values of APF and RPC. This is exemplarily visualized in Figure. 4.6 with different values of α applied to the same data set. The vertical dashed lines additionally highlight the discovered patterns with varied α values. As shown in the figure, the patterns having low RPC values are, obviously, close to zero. Hence, the distance between zero and the original data values is already low. If more pattens have low RPC values, due to high values of α, that leads to decreasing values of DM^k, even though the RDC did not increase its actual quality. A solution to this special problem may be the normalization of every RPC value by the original data:

$$RPC_j^{norm} = \frac{RPC_j}{RPC_j^{orig}} \tag{4.5}$$

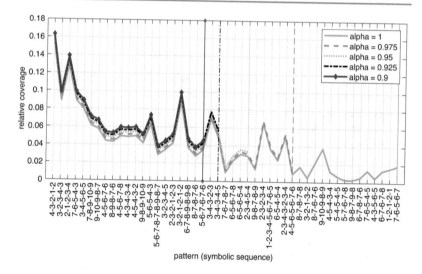

Figure 4.6 Comparison of RPC-spectra with varying pattern discovery parameters α

The normalized RPCs of all RDCs as well as the original data are exemplarily visualized in Fig. 4.5c. The adapted formula for the normalized L_k norm is:

$$DM_{norm}^k = \frac{\left(\sum_j \left| RPC_j^{norm} - 1 \right|^k \right)^{1/k}}{num\,Patterns} \qquad (4.6)$$

Beside the distance measure, it is desired to establish a similarity measure that can be used to express the quality of an RDC between 0 and 1. A well-defined quality measure should cover the whole range. Due to the fact, that the RPC-spectrum has values between zero and one it is guaranteed for $k \geq 1$ that DM^k will lead to values in the desired range. Hence, a similarity measure can be defined as:

$$SM^{k\geq 1} = 1 - DM^k \qquad (4.7)$$

$$SM_{norm}^{k\geq 1} = 1 - DM_{norm}^k \qquad (4.8)$$

For $k < 1$ the distance measure will likely output values higher than one. In this case, a similarity measure within the desired range can not be created.

Table 4.1 Values of both proposed quality measures based on the Euclidean distance for the synthetic test case of Fig. 4.5

	RDC 1	RDC 2	RDC 3
SM^2	0.9991	0.9995	0.9941
SM^2_{norm}	0.9743	0.9869	0.8519

In Table 4.1 the similarity between the RDCs and the original data is exemplarily shown for the Euclidean distance as well as for the normalized Euclidean distance. The values for RDC 3 are the lowest in both measurements, as we designed it in this synthetic test case. Unfortunately, the SM^2 values of all RDCs are very close to 1, even though RDC 3 was designed as a bad RDC. This is caused by the curse of dimensionality as well as high α values. In later evaluations of Section 5.2 the practicability of L_k norm with varying k as well as the normalization of RPC values will be evaluated in more detail.

4.1.4   Influence of Pattern Discovery

In previous sections APF and RPC were proposed as unsupervised CPs for the RDC construction. At this point it is necessary to clarify that these CPs are neither totally unsupervised nor completely independent from expert knowledge. This is because the CPs are influenced by the pattern discovery settings. Depending on these settings the method discovers different sets of patterns, that vary e.g. in the total number of patterns, their length and level of detail. Hence, the CPs also vary with the change of settings.

Nevertheless, this is not a disadvantage of the approach, but a necessity. The pattern-based approach for RDC construction is made to be a generalized approach, easy to transfer to new use cases and requirements by the use of the unsupervised pattern discovery. With every generalized approach, there is the need for adaption to individual requirements. These individual requirements can be captured by adapting the pattern discovery settings. In general, the discovered patterns need to cover the level of detail, that is necessary for the application. For example an RDC, that is supposed to be representative for the driving behavior regarding the ratio of urban, suburban and highway driving, does need less detail than another RDC that is supposed to be representative for the driving style of comfortable or sporty drivers. In the latter RDC the patterns also need to capture e.g. the gradient of velocity or the longitudinal and transversal acceleration. This level of detail may be to high for the first example, due to the fact that it only needs to capture the rough value of the vehicles velocity. The user can influence that level of detail by choice of:

- Signals to be included in the pattern discovery
- Number and location of discretization boundaries
- Parameters of (logarithmic) symbolic reduction

In Figure. 4.7 this influence is exemplarily shown by the differentiation between acceleration events with varying settings. The pattern discovery was applied three times with different settings to one and the same vehicle data set. In Figure. 4.7a set-

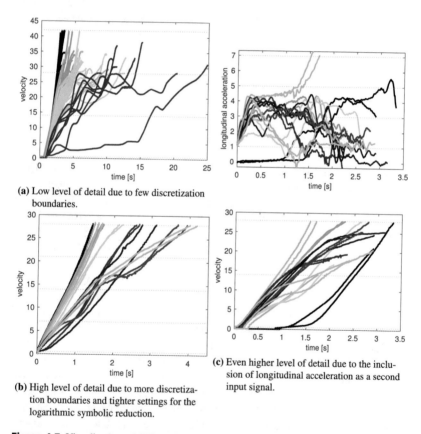

(a) Low level of detail due to few discretization boundaries.

(b) High level of detail due to more discretization boundaries and tighter settings for the logarithmic symbolic reduction.

(c) Even higher level of detail due to the inclusion of longitudinal acceleration as a second input signal.

Figure 4.7 Visualization of different acceleration events between 0 and approximately 30 km/h to highlight the influence of pattern discovery settings regarding the level of detail of discovered patterns. Three different settings were applied to the same data set. For each setting separately members of the same pattern are shown in the same color

tings were chosen to discover patterns with only low level of detail. The horizontal dashed lines show the corresponding discretization boundaries. Five different patterns are plotted that show different kinds of acceleration events. In Fig. 4.7b more detail was added by the use of more discretization boundaries and tighter settings for the logarithmic symbolic reduction. It shows the differentiation of acceleration events in much more detail. Additionally, Fig. 4.7c shows the inclusion of a second signal, the longitudinal acceleration. It enables the pattern discovery to differentiate sequences in even more detail concerning the gradient of velocity.

4.2   Identification of Full Trip RDC

This section deals with the implementation of a pattern-based full trip method. As already reviewed in Section 2.2.3, full trip methods are relatively simple approaches to identify RDCs from a set of existing trips. Hence, the RDCs are trips that were actually driven in real life. In the literature the most common approach is to calculate knowledge-based CPs and then select the trip that is either the closest to the center of all trips or has the smallest distance to the parameter values of the original data.

The proposed pattern-based full trip approach solves the problem of finding accurate CPs, that match the actual goal of the application. This task is done by the pattern discovery. The user only has to ensure, that the method discovers patterns, which capture the necessary level of detail by choice of included signals and pattern discovery settings. In most cases this task is much easier than developing CPs and a corresponding function. Imagine an RDC that is required to be representative for the energy consumption of a vehicle. In knowledge-based approaches the necessary CP, the energy consumption, first needs to be modeled for the individual vehicle including the relevant powertrain components, which is more complex than fixing the pattern discovery settings. In most cases only two parameters need to be fixed by the user for the pattern discovery, namely the number of bins ($noSteps$) and the basis of symbolic reduction ($logBasis$). In practice, default values for all the other parameters already provide good results. The influence of pattern discovery settings was already discussed in Section 4.1.4.

In general, the required pieces to implement a pattern-based full trip method were already introduced in Section 4.1. In Section 4.1.1 pattern-based CPs were introduced by means of APF and RPC. Their usage in terms of identifying a full trip RDC was also briefly described in use case 3 of Section 4.1.2. Even the basic distance and similarity calculation between RDC and original data was introduced in Section 4.1.3 as a quality measure. Nevertheless, for better understanding it is necessary to explain the approach in more detail. Therefore, in the following sections the process of cutting the data into trips and subsequently obtaining a set of patterns for every trip is described. Furthermore, the selection process of an RDC is described and discussed.

4.2.1 Trips and Pattern Sets

From Section 2.2.3 the definition of micro trips is already known. In most cases the data is cut at zero velocity to obtain micro trips. Unfortunately, this approach is not suitable for the pattern-based identification of RDCs, which is evident looking at the following example: A vehicle driving in a stop and go situation in a traffic jam will experience many short micro trips showing the acceleration from zero to small velocities followed by a deceleration to zero. If these micro trips were the input to the RDC identification, a majority of those micro trips could be represented by just one pattern, the acceleration and immediate deceleration. But the RDC aims for the representation of all the possible situations. Hence, an RDC should represent the stop and go situations as well as all the other situations. This is why micro trips in general are not suitable. It is preferred to input trips that are longer, while the beginning and ending still should be at zero velocity. Beyond that, a trip should be completed in a way that excludes driving of the vehicle immediately before and after that cut. It is recommended to cut the data into a new trip, when there is a considerably long break with a duration of at least 15 minutes. A trip therefore includes data for e.g. the drive to work. This cutting scheme is exemplarily shown in Figure. 4.8. Every highlighted area defines a separate trip.

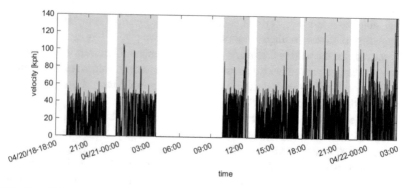

Figure 4.8 Cut data into trips when there is a break of at least 15 minutes based on an exemplary data set of a Volkswagen ridesharing vehicle

For the selection of an RDC from all trips based on patterns it is necessary to prepare the same kind of pattern set for every trip. If this is not the case, the APF- or RPC-spectra from different trips can not be compared. Hence, even if a pattern does not exist in a trip, there needs to be a record. To ensure this, there are basically two different approaches for the implementation of pattern discovery. First, it is possible to apply the pattern discovery separately to every trip and then align every pattern set of every trip. Secondly, the pattern discovery can be applied to the large data set that contains every trip. This global set of patterns then needs to be divided regarding each trip.

4.2.2   Selection of RDC

The selection of an RDC from a set of trips is done by assessing every trip regarding the distance measure introduced in Section 4.1.3. In general, there are multiple options for the selection based on the distance measure as visualized in Table 4.2. These options require the formula of the distance measure to be slightly adapted. The adaptions are shown in the last row of the table. While previously only the possibility to use the RPC as a characteristic parameter was mentioned, in two of three options it is now possible to also use the APF. This is why RPC in these formulas is replaced by the general term CP.

First of all, it is possible to select the trip that is closest to the center based on either the APF or the RPC. The center μ_j is calculated by averaging the CP values of every trip i in each dimension j separately:

$$\mu_j = \frac{\sum_i CP_{j,i}}{num\,Patterns} \tag{4.9}$$

Subsequently the distances between every trip i and μ_j can be calculated by the L_k norm. The trip with the lowest distance is selected to be the RDC.

Another possibility is to calculate the trips distances to the original data in total. Hence, the RPC-spectrum of the original data before cutting it into trips is calculated. This was already proposed in Section 4.1.3. Note that it is not possible to use the APF in this case, because the APF-spectrum contains the cumulated absolute pattern frequencies of all trips, which is not comparable to the trips APF-spectra. The third possibility is to calculate the distances of every trip to every other trip and then select the trip that has the least average distance to all trips. Therefore, the formula is adapted to calculate the distances between two trips i_1 and i_2.

Table 4.2 Possibilities to select an RDC from a set of trips

	Closest to center	Least dist. to original data	Least average dist. to all trips
APF	✓	✗	✓
RPC	✓	✓	✓
Formula	$\dfrac{\left(\sum_j \left\lvert CP_{j,i} - \mu_j \right\rvert^k\right)^{1/k}}{num\,Patterns}$	$\dfrac{\left(\sum_j \left\lvert CP_{j,i} - CP_j^{orig} \right\rvert^k\right)^{1/k}}{num\,Patterns}$	$\dfrac{\left(\sum_j \left\lvert CP_{j,i_1} - CP_{j,i_2} \right\rvert^k\right)^{1/k}}{num\,Patterns}$

While these three approaches in practice will only show minor differences in results, it is not guaranteed to get the same results. Comparing the approach of column one (closest to center) and two (least dist. to original data) of Table 4.2, it can be shown that the CP values of the column two approach will be weighted by the trips lengths. Hence, values of longer trips have a higher impact on the selection process. Using the column one approach, the characteristic parameter values are implicitly normalized with regard to their length.

Furthermore, the pairwise distances of the column three approach in Table 4.2 can also be used as a basis for clustering algorithms. In this case the set of all trips can be separated into similar groups of trips. Subsequently an RDC for every group can be extracted. This is valuable especially in applications with a high diversity of trips. However, this work focuses on constructing the best RDC, which is why the possibility of clustering will not be handled in the following sections. Nevertheless, every idea that is proposed to construct one RDC can easily be extended by a preceding clustering algorithm.

In Equations (4.5) and (4.6) the possibility to normalize the CPs was already introduced to cope with decreasing values of DM^k when increasing α. In this case the values are normalized by the CP values of the original data. In general, this normalization aims to create CP values that cover the whole range between 0 and 1, which is not guaranteed when using the original data as a reference. This is why the formula to calculate the normalized CPs is adapted by including the statistics of all trips:

$$CP_j^{norm} = \frac{CP_j}{\mu_j + 3 * \sigma_j} \tag{4.10}$$

In equivalence to μ_j, the variable σ_j describes the standard deviation. In doing so, the normalized values have a higher probability to be in the range of 0 to 1, while at the same time outliers are ignored.

4.3   Construction of RDC

The idea of pattern-based RDC construction includes the identification of all patterns that in total represent the majority of the data. Then a much smaller set of patterns has to be selected that still is representative for the data. All patterns from this representative pattern set need to be concatenated in a reasonable order. This approach has the potential of constructing the shortest possible RDC, while maintaining the representativeness for the original data. At the same time a high flexibility is given regarding the choice of representative patterns. Hence, the representativeness by means of the pattern set can be adapted to various requirements. Different use cases were already discussed in Section 4.1.2. The approach, again, includes all the advantages resulting from the unsupervised pattern discovery, especially the ability to analyze large amounts of data.

While the pattern-based identification of RDCs, introduced in Section 4.2, had the advantage of simplicity, the pattern-based construction is much more complex. The biggest challenge is the construction of RDCs that are technically correct. Hence, they need to be realistic and drivable for a vehicle. When concatenating patterns in order to create the RDC, the technical correctness places high demands on the transitions between patterns. Due to the complexity, this section limits its focus to the univariate case of only taking the velocity of vehicles into account.

In Section 4.3.1 the selection of representative pattern sets is described. Based on these sets Section 4.3.2 proposes a transformation of the pattern order problem into a directed graph problem, in order to find the best possible sequence of patterns in the RDC. Finally, Section 4.3.3 deals with the concatenation of patterns and the design of transitions to construct technically correct RDCs.

4.3.1   Selection of Absolute Pattern Frequency

The first task to construct an RDC by concatenating patterns is the selection of a representative set of patterns, that will subsequently be used in the RDC. As described in Section 4.1.2 by means of use cases 1 (evenly distributed frequency), 2 (representative frequency) and 4 (average trip), there are multiple techniques to determine such a set of patterns and, thus, to create a desired APF-spectrum for the RDC. This pattern set could theoretically be used directly for the construction, but due to the characteristic of the pattern discovery algorithm to identify partly or fully overlapping patterns, it is recommended to slightly correct the pattern set.

The problem gets more clear with the following example, which is visualized in Figure. 4.10. The pattern discovery is applied to a data set and aims to cover at least 90 % of it. In one of the earlier iterations of postprocessing the method selects the pattern with the symbolic sequence 7-8-7-8-9 that occurs 19 times. In order to exceed the 90 %, the pattern 7-8-7-8-7-8-9-10-8-7 is additionally selected in a later iteration, which occurs 4 times. The APF-spectrum is shown in Fig. 4.10a. Unfortunately, the symbolic sequence of the earlier chosen pattern is completely covered in the later chosen pattern, but only in 4 of 19 cases as visualized in Fig. 4.10c. The time series named *dim1* is shown as dashed line, while members of the earlier pattern are highlighted in gray and the later pattern in black. An example of an occurrence of both patterns is additionally shown in Fig. 4.10b. Clearly, the black pattern includes the gray pattern in this case. The same incident can be observed with patterns 6-5-6-7-8 and 8-7-6-5-6-7-8-9-10.

When constructing an RDC it is recommended to correct the APF according to those fully overlapping patterns. If it was not corrected, in this example the RDC would contain the representative amount of pattern 7-8-7-8-9 added by the representative amount of pattern 7-8-7-8-7-8-9-10-8-7. In total, the pattern 7-8-7-8-9 would be overrepresented. The original APF is corrected by subtracting the APF of the overlapped pattern by the APF of the overlapping pattern. The resulting corrected APF is shown in Fig. 4.10a as dashed line. In fact, there is the possibility of discovering a set of patterns that contains a pattern, that is included in a bigger pattern, that is included in an even bigger pattern and so on. These special cases need to be considered, when correcting the APF.

4.3.2 Solving a Directed Graph Problem

After selecting a representative set of patterns, the problem of choosing the optimal order to concatenate the patterns has to be solved. As already mentioned, the biggest challenge in constructing an RDC by concatenating patterns is the technical correctness. In this step it means, that the ending of a pattern needs to match the start of the next pattern in order to create a smooth transition. This section proposes the transformation of the problem into a graph problem, more specifically an Asymmetric Traveling Salesman Problem (ATSP). The basic idea is to model patterns as nodes in the graph and the edges between patterns as the velocity difference between the last value of a pattern and the first value of the next pattern. The determination of the exact solution to the ATSP is computationally heavy. Nevertheless there are multiple heuristic algorithms that are able to find good approximate solutions.

Asymmetric Traveling Salesman Problem
The Traveling Salesman Problem (TSP) describes the problem of a salesman having a given number of places, that he or she has to visit. The target is to find the shortest possible route that includes all the places exactly once and ends in the same place that

the salesman started from. Hence, the corresponding cost function is the accumulated distance of all trips. The TSP can be pictures as a Graph with every place being a node and the trips between two places being an edge with a weight that defines the distance between both places. In theory, every node is connected with every other node, which results in a large set of solutions. The calculation of the exact solution is classified as NP-hard. While the TSP can be modeled as a graph with undirected edges, the asymmetric TSP (ATSP) has to be modeled by means of directed edges. In case of the ATSP the distance between node A and B is not equal to the distance between node B and A. Hence, the edges between both nodes have not the same weight.

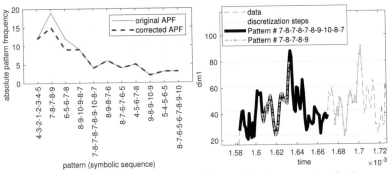

(a) Original and corrected APF-spectrum due to fully overlapping patterns.

(b) Zoomed into time series shown in (c).

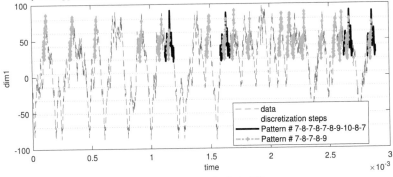

(c) Time Series with two exemplary patterns that overlap in 4 of 19 cases.

Figure 4.9 Example of fully overlapping patterns and the proposal of correcting the APF before constructing the RDC

Transformation into Graph Problem

The transformation of the pattern order problem into a ATSP problem is done by modeling the patterns as nodes and the edges weights as delta of last and first values of preceding and following pattern. The solution of the ATSP subsequently is the best order of patterns with the least jumps between patterns. The basis is a representative set of patterns and their frequency. The selection process was already described in Section 4.3.1. The transformation is shown in Figure. 4.9 by a minimal example. In Fig. 4.9a (left) three different patterns are shown, that were previously chosen to be in the RDC. Pattern 1 and 3 need to occur twice in the RDC, while pattern 2 is needed only once. On the right side of this figure a reasonable order of those patterns is already shown. In this simple case it may be easy to find the best order for concatenation, but the more patterns there are the harder the selection is.

The graph representation of this minimal example is pictured in Fig. 4.9b. One node for every pattern member that needs to be included is necessary for the transformation. This is, because the formal ATSP says that every node is only allowed to be visited once. The graph shows two edges between every pair of nodes with their corresponding weights. Looking at pattern 1 and 2, the edge from pattern 1 to 2 has zero weight, because the velocity delta of the last value of pattern 1 ($30\ km/h$) and the first value of pattern 2 ($30\ km/h$) is zero. In the other direction there is a delta of $30\ km/h$ between the last value of pattern 2 and the first value of pattern 1. As shown in the figure, there are two more nodes s and e that are placed as dummy nodes. Node s marks the beginning of an RDC. The RDC needs to start at zero velocity, which is why there are only edges from s to patterns whose first values are at zero velocity. Due to the requirement, that an RDC also needs to end at zero velocity, the same technique is applied to node e, that marks the end of the RDC. Furthermore, the ATSP says that the salesman should end at the same place that he or she started from. This is why an edge needs to be added between nodes e and s. The resulting graph can now be solved regarding the ATSP by common algorithms that are described later in this section. The solution is the optimal order of patterns with the least value jumps between concatenated patterns. Hence, the pattern order for the RDC is now fixed.

In case of the designed example the transitions between patterns matched perfectly, which is why many edges have zero weight, but usually this is not the case. Inaccuracies need to be considered when implementing such an RDC construction method. First of all, members of a pattern do not necessarily have exactly the same first and last values. This is why the weight of edges represents the delta of median first and last values of all members. Secondly, the transitions may not be as perfect as shown. Additional smoothing functions need to be applied in order to optimize the transitions, which is described in more detail in Section 4.3.3.

Solution of ATSP

In general, TSPs and ATSPs can be described as binary integer programs. Hence, algorithms solving these problems need to decide if an edge is part of the solution (1) or not (0). The objective function in this case describes the accumulated weights of all used edges. The minimization can be performed by linear programming (LP) techniques, due to the linear dependency from the selected edges. These LP techniques were originally developed for problems in the field of *Operations Research*. The history of exact and heuristic approaches to solve linear problems is reviewed in [96] especially for the TSP. In general, the TSP is not a complex problem as it can easily be mathematically formulated. Unfortunately, in accordance with NP-hard problems the computation time grows fast with increasing number of nodes in the graph. Consequently, the number of possible routes result to 181,440 when using only 10 nodes. With 17 nodes there already are approximately 10^{13} possible routes. Because in this application of RDC construction there will usually be even more patterns, it is necessary to use heuristic approaches. Those heuristic approaches aim to find approximate solutions that are close to the optimal solution. The Branch and Bound approach is commonly used to solve the TSP and other LP problems. The first algorithmic implementation was published in 1965 in [15]. It describes an approach of splitting the large solution space into smaller pieces, so called sub-problems. In doing so, a tree structure can be extracted that divides the problem into smaller sub-problems with every new leaf being an even smaller problem. At some point the sub-problems are small enough to calculate exact solutions. This is done by the simplex approach proposed in [68], which is another commonly used technique in LP problems.

Translated back to the pattern order problem, the large graph of nodes is divided into smaller graphs and subsequently solved by simplex algorithm. Hence, these sub-solutions are small groups of patterns as well as their optimal order within this small group, so called sub-tours. After the solution of those sub-problems, it is necessary to connect the sub-tours with each other in the optimal way. Because the initial problem was already solved to a large extent, the connection of sub-tours is now much less computationally heavy.

Graph Weights

On the basis of Figure. 4.9 the transformation of the pattern order problem into a graph problem including a technique to determine the edge weights was already introduced. Additionally, this section now introduces an alternative technique. As already described in Section 4.3.1, the pattern discovery results will eventually contain patterns that fully overlap. Within the selection process of a representative

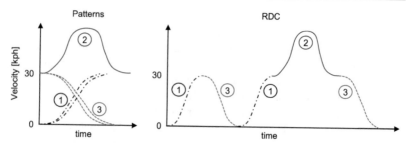

(a) Left: Exemplary set of representative patterns to be included into the RDC. Right: Exemplary
solution of optimal pattern order.

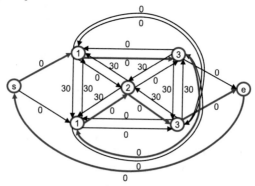

(b) Corresponding graph visualization of example in (a) with additional starting and ending dummy
nodes (s, e). The optimal path through the graph that solves the ATSP is highlighted with thick
lines.

Figure 4.10 Exemplary transformation of pattern order problem into a graph prolem

pattern set these cases are already considered and corrected. But beside these fully
overlapping patterns, there may also be partly overlapping patterns. The handling of
these cases is exemplarily shown in Figure. 4.11 by means of an alternative weight
calculation.

In Fig. 4.11a two exemplary patterns with all their members are visualized in
different colors. Their symbolic sequences are shown in the legend. The original
data was synthesized by a random walk. When analyzing the symbolic sequences of
both patterns, it is conspicuous that the end of pattern 6-5-4-5-6 (pattern A) and the
beginning of pattern 4-5-6-7-8-9-8 (pattern B) is equal in 3 symbols, namely 4-5-6.
As shown in Fig. 4.11b some occurrences of both patterns have in fact an overlap.
Hence, when concatenating the patterns in the RDC it may not be ideal to connect

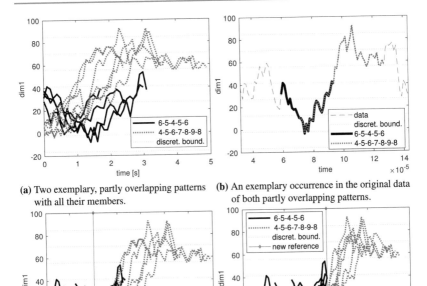

(a) Two exemplary, partly overlapping patterns with all their members.

(b) An exemplary occurrence in the original data of both partly overlapping patterns.

(c) Visualization of both patterns, but shifted according to the overlap in the solid-drawn pattern. Every sample on the vertical line is considered in the weight calculation. The solid-drawn pattern is considered to be shortened, while the dashed-drawn pattern is considered to be used to full extent.

(d) Visualization of both patterns, but shifted according to the overlap in the dashed-drawn pattern. Every sample on the vertical line is considered in the weight calculation. The dashed-drawn pattern is considered to be shortened, while the solid-drawn pattern is considered to be used to full extent.

Figure 4.11 Exemplary visualization of partly overlapping patterns and the handling by means of weight calculation

them end to end, due to the high delta of last and first values. Instead, it would be much better to concatenate them with an overlap according to the equality of symbols 4-5-6. Beside this, the approach has the potential of increasing the representativeness of the RDC. Imagine a representative set of patterns with one occurrence of each pattern A and B, among other patterns. If all patterns are concatenated end to end, the sequence 4-5-6 is overrepresented. This overrepresentation can be reduced by considering an overlap when constructing the RDC.

In order to implement the concept of overlapping patterns in the concatenation process, described in the following Section 4.3.3, there is the need to adapt the calculation of weights in the graph. While previously the weight was calculated by the median of last values (of pattern A) and first values (of pattern B), now the overlap also needs to be considered. This technique is exemplarily shown in Figs. 4.11c and 4.11d. In general, the approach is to use either one of the patterns to full extent and cut the other one. In Fig. 4.11c the distance is calculated between the clipped last values of pattern A and the original first values of pattern B. The red vertical line shows the corresponding values of both patterns that are considered for the distance calculation. In Fig. 4.11d the other possibility is shown. The distance is calculated between the original last values of pattern A and the clipped first values of pattern B. To set the weight of the edge between the corresponding nodes, the minimum distance of both possibilities is calculated.

4.3.3   Design of Transitions between Patterns

After solving the directed graph problem and consequently the pattern order problem, the patterns now only need to be concatenated in the determined order. Unfortunately, in order to construct a technically correct RDC, there are a few more challenges to overcome. First, possible value jumps at the transitions between patterns in the RDC need to be avoided. Even though the previous section introduced a solution to find the best order of patterns, this does not necessarily mean, that consecutive patterns perfectly match. Secondly, the actual members that are inserted in the RDC are not yet selected. Until now only the order of patterns is solved. Thirdly, beside absolute value jumps, abrupt changes in the gradient of the velocity need to be avoided.

Fill Gaps

To avoid absolute value jumps in the transition between patterns, the general approach is to fill these gaps with other patterns. In this case a gap is defined as a symbolic value jump that is greater than one. An example is visualized in Fig. 4.12a. The dotted-lined pattern was previously chosen to follow the solid-lined pattern in the RDC. Even though the pattern order problem was solved, there is a jump between those two patterns. The symbolic sequences of both patterns show a difference of two symbols. The true delta in velocity varies depending on the chosen members. When a gap is identified, another pattern is searched in the dictionary that matches the last symbol of the preceding pattern as well as the first symbol of the following pattern with a symbolic distance of at most one. In Fig. 4.12b a matching pattern is shown in between both patterns. Note that potentially all members of the gap closing pattern could be considered. In this figure only one exemplary member is shown. Unfortunately, these additional patterns reduce the representativeness, because the APF- and RPC-spectra are manipulated and, thus, they differ from the optimal ones. Nevertheless, these adaptions are necessary to construct a technically correct RDC.

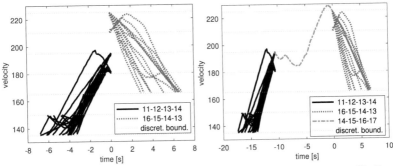

(a) Two consecutive pattern with all their members.   (b) The gap between both patterns was filled by an additional pattern.

Figure 4.12 Exemplary case with the necessity to fill the gap between two patterns

Choose Pattern Members

In general, the choice of members could simply be made by going through all the patterns chronologically according to the pattern order and selecting the member that matches the best with the previous pattern. Unfortunately, the current best match with the previous pattern does not necessarily lead to a good match with the following pattern. This is why in this section a two step approach is introduced for the choice of pattern members. In the first step the variety of members for each pattern in the RDC is reduced to those, who at least have an acceptable match. In a second step, the best match is inserted into the RDC.

First, another necessary condition is proposed: The velocity gap between two concatenated patterns has to be smaller than the size of a bin, which is defined by the distance between two consecutive discretization boundaries. Even though the symbolic gaps have already been filled with additional patterns in the previous section, there still might be high value jumps. This is exemplarily visualized in Figure. 4.13.

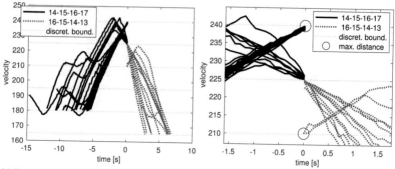

(a) Two consecutive pattern with all their mem-   (b) Zoom window to show details of the trans-
bers.   ition.

Figure 4.13 Visualization of two consecutive patterns and the value jumps between their members. Even though they symbolically match, it is possible that two members have a high velocity distance

In Fig. 4.13a two patterns with all their members are shown. In theory, every member of the earlier pattern could be combined with every member of the later pattern. Fig. 4.13b shows a small window of these patterns that captures the transition. It is evident, that there are members of both patterns matching nearly perfectly. Beside this, there are also members that have high distances. The corresponding values are encircled in the figure. These members are not suited to be connected, which is why they need to be avoided. As introduced, the distance limit is set to one bin size. The encircled members, therefore, can not be considered for concatenation.

The basic question to answer here is: In order to reach the end of the RDC without gaps, which member has to be selected in the beginning? Hence, a set of members is needed for every pattern in the pattern order with every member having acceptable distances to the preceding and following patterns. This reduction of pattern members is exemplarily visualized in Figure. 4.14a. Imagine four patterns to be included in the RDC in a fixed order. The black boxes in the upper part of the figure each represent a member of the according pattern. Hence, pattern 1 has three members, pattern 2 has two members and so on. In theory, every possible member could be connected in the RDC by every other member of the preceding as well as the following pattern, which is represented by connections between the boxes. In the lower part of the figure the reduction process is shown. It begins with the last pattern and considers every member to be part of the reduced set. The dashed connections, in contrast to the solid connections, show velocity jumps between those members that are greater than the bin size. The first member of pattern 3 is highlighted as

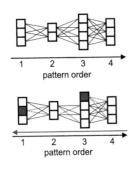

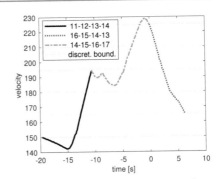

(a) Reducing the set of possible members.

(b) An exemplary result of three patterns and their chosen members to construct the RDC.

Figure 4.14 Selection of a member from each pattern in the pattern order

red box, because there exists no connection to pattern 4 with an acceptable value jump. Red boxes are not part of the reduced member set. Because this member is eliminated, its connections to the pattern 2 members are not considered anymore. In the transition between pattern 2 and 3 there are single combinations that have high value jumps, but no member needs to be excluded from the reduced member set. In the last transition between pattern 1 and 2, again one member can be classified as unsuitable, because all of its connections to pattern 2 members are above the distance limit. By this backwards iterative approach a set of members is created that leads to acceptable transitions in all the following patterns.

If there is a pattern without any members in the reduced set, another pattern with a better match has to be included. This process is equivalent to the filling of gaps in previous section. Finally, based on the reduced member set a best match is selected for every pattern. An example for three matching members is shown in Figure.4.14b. It is related to the example in Figure.4.12.

Smoothing of Transitions
The previous descriptions of the design of transitions focused on absolute value jumps, but the gradient of velocity also needs to approximately match. In Figure.4.14b the first transition shows an abrupt change of the gradient, because the patterns were just concatenated end to end. Those abrupt changes are technically incorrect due to e.g. inertia of the vehicle. This section proposes an approach to slightly shift the point of concatenation of both patterns in order to construct less abrupt transitions. For every transition a fixed window of potential shifting is defined

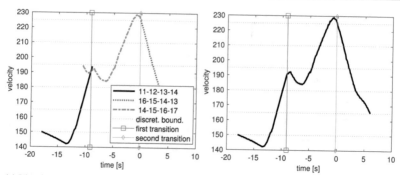

(a) Visualization of three patterns shifted by the optimal amount of samples to create the smoothest transitions.

(b) The resulting RDC with smooth transitions.

Figure 4.15 Smoothing of transitions by considering an overlap window to find the best matching samples in both patterns regarding the absolute value as well as the gradient

for every pattern. In Figure. 4.15a an example is shown with a shifting window size of 20 samples. Within the last 20 samples of the earlier pattern and the first 20 samples of the later pattern the best transition is determined by pairwise analysis of the absolute value as well as the gradient. In this example the first pattern was cut by the last two samples. The second pattern was cut by the first 14 samples. The new point of concatenation is highlighted by the red vertical line. Additionally in this figure, another transition is shown that in general matches much better. Hence, less samples needed to be cut away. In Fig. 4.15b the resulting RDC is shown. In exceptional cases with the overlap being not too successful, it may be necessary to additionally smooth the transition e.g. by a moving average.

Evaluation

5

In previous sections algorithms were proposed for unsupervised pattern discovery in multivariate time series (Chapter 3) and applied for the construction of pattern-based RDCs (Chapter 4). This section aims to validate and evaluate the proposed algorithms and approaches. First, Section 5.1 deals with the pattern discovery itself independently of the use case of RDC construction. In Section 5.2 follows the evaluation of pattern-based RDC construction approaches in contrast to common knowledge-based techniques.

5.1 Experimental Evaluation of Unsupervised Pattern Discovery

A major part of this section (Sections 5.1.1 to 5.1.3) was already published in previous work of [71] and is now contextualized to the whole framework. The pattern discovery framework will be extensively tested in four experiments. The experiments aim to show the benefit of the proposed preprocessing methods *hysteresis*, *unification* and *logarithmic reduction* as well as the *greedy postprocessing* method in comparison to common methods. Section 5.1.1 contains a parameter experiment based on a two-dimensional synthetic data set. Additionally, multiple quality criteria are introduced to evaluate the quality of patterns as well as the quality of parameter settings based on every discovered pattern. By performing grid search the quality of parameter settings is evaluated and discussed. This section especially focuses on the evaluation of filtering approaches including the hysteresis. Section 5.1.2 proposes an experiment based on a real-life data set with > 100 mio. samples. By discovering patterns in the univariate time series of a battery-electric-vehicle, the benefit of logarithmic reduction in contrast to standard reduction is shown. Beside this, the performance of the pattern discovery framework in large data sets is demonstrated.

F. K. D. Noering, *Unsupervised Pattern Discovery in Automotive Time Series*,
AutoUni – Schriftenreihe 159, https://doi.org/10.1007/978-3-658-36336-9_5

Section 5.1.3 additionally underlines the performance of the unification in an experiment with increasing dimensionality. It is based on a synthetic five-dimensional time series. Last but not least, Section 5.1.4 evaluates the effectiveness of postprocessing techniques to reduce the amount of patterns while at the same time maintaining the relevance of discovered patterns. This experiment is based on the synthetic data set from Section 5.1.1. All tests are done on single cores of a workstation with two Intel™Xeon™Gold 6134 CPUs ($3.20GHz$).

5.1.1 Synthetic Randomized Data Set

Experiment Description
The first test case for the pattern discovery is based on synthetic data and the testing framework of previous work in [70] with just small changes. In general, it includes the construction of a data set with random patterns as well as random fluctuation in between the patterns, the application of the pattern discovery with multiple parameter sets and finally the evaluation via Jaccard index. The data set consists of many different patterns which are created by applying a random walk with random step size and random number of steps. Beyond that, the random walk is modified by setting limits for minimum and maximum values. A varying number of copies for each pattern is generated by duplicating it as well as adding some distortion (noise and longitudinal scaling). Finally, the patterns and its noisy copies are concatenated in a random order to create the test data set. In previous work the patterns were concatenated by inserting a noisy linear function between the last sample of the previous pattern and the first sample of the following pattern. In this work we replaced the linear function by the modified random walk, which runs until the current sample is in a similar value region as the first sample of the following pattern. In doing so, more randomness is added to the time series. Furthermore, the framework is adapted to create multivariate patterns, meaning the modified random walk runs for each dimension independently but with the same number of steps. Hence, a pattern contains a sequence of univariate samples in every dimension with the same number of samples for every dimension. Beside these two adaptions, more patterns per data set are created to challenge the methods even more. However, this is also why the following results are not exactly comparable to the results of previous work [70]. In addition to the Jaccard index, which is usually called *overlap*, the *True Positive Rate* (TPR, also known as recall) as well as the *False Discovery Rate* (FDR) is added to the focus of this work. These three characteristics are defined as:

$$overlap = \frac{|TP|}{|TP \cup FN \cup FP|} = \frac{|TP|}{|I_{pre} \cup I_{disc}|} \tag{5.1}$$

$$TPR = \frac{|TP|}{|TP \cup FN|} = \frac{|TP|}{|I_{pre}|} \tag{5.2}$$

$$FDR = \frac{|FP|}{|TP \cup FP|} = \frac{|FP|}{|I_{disc}|} \tag{5.3}$$

$$\text{with: } TP = I_{pre} \cap I_{disc}$$
$$FP = I_{disc} \setminus I_{pre}$$
$$FN = I_{pre} \setminus I_{disc}$$

While comparing one predefined pattern with one discovered pattern, TP (true positive) comprises all the indices which are both in the predefined I_{pre} and in the discovered pattern I_{disc}. FP (false positive) are those indices that are in the discovered pattern but not in the predefined pattern, while FN (false negative) are those that are part of the predefined pattern but not of the discovered pattern. In Figure 5.1 these index sets are exemplarily visualized for one member of a predefined and a discovered pattern. The dashed line shows the time series, while the upper solid line is the predefined pattern and the lower solid line is the discovered pattern. In fact, the members of a pattern are not handled separately as shown in Figure 5.1, but simultaneously. Hence, I_{disc} for example contains all the indices of all members of one discovered pattern. An overlap score of 0 describes a mismatch. A score of 1 describes a perfect match between predefined and discovered pattern. Note that the overlap also decreases from a perfect match if the detected pattern is longer than the predefined one. The calculation is performed for every combination of predefined and detected pattern. Afterwords a best match for every predefined pattern can be chosen. By calculating the best matches and averaging their evaluation characteristics, we are able to express the performance of a parameter setting.

This experiment aims to find the best performing parameters concerning the characteristics overlap, TPR, FDR and runtime within the parameter space, which is shown in Table 5.1. By performing grid search, more than 2000 different parameter settings have been evaluated including settings for the baseline approach. For every parameter combination the same synthetic time series is tested, which has roughly 700,000 samples in each of two dimensions and 200 randomly created patterns having between 10 and 30 repetitions.

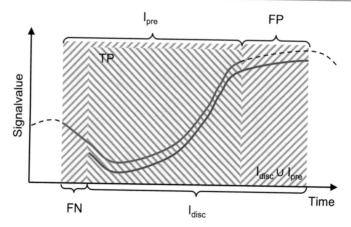

Figure 5.1 Exemplary visualization of index sets for the calculation of overlap, TPR and FDR[1]

Results

Before going into detail, for visual validation in Figure 5.2 nine exemplary patterns with all their members are shown, which were discovered in a synthetic data set with two dimensions. The dashed subsequences correspond to the first dimension, while the solid ones belong to the second dimension. To preprocess the time series, all the proposed techniques were used, namely hysteresis, unification and logarithmic reduction. The method is able to discover visually valid patterns in a two-dimensional space with variable length.

As described earlier in this section, a brute force experiment is performed to find the best parameter settings. In Figure 5.3 the results for every parameter setting concerning the overlap, TPR, FDR and runtime are shown. A good parameter setting should result in a high overlap, low runtime, high TPR and low FDR. In general, the pattern enumeration algorithm dominates the adapted Sequitur algorithm in every evaluation characteristic. But note, that independently of the pattern discovery algorithm, the result can be highly improved by adding the proposed preprocessing steps. This becomes clear when comparing the results with the baseline results that are highlighted in the figure. For further evaluation this work focuses on the pattern enumeration algorithm, because of its dominance.

---

[1] "Intelligent Data Analysis, vol. 25, no. 5, Noering et al., Improving Discretization Based Pattern Discovery for Multivariate Time Series by Additional Preprocessing, pp. 1051-1072, (2021)", with permission from IOS Press. The publication is available at IOS Press through http://dx.doi.org/10.3233/IDA-205329.

Table 5.1 Parameter space of experiment

Section	Method	Parameter	Parameter values
Section 3.3.1	PAA	windowsize	10, 20, 30, 40
		stepsize	windowsize/2
Section 3.3.1	Wavelet	denoiselevel	1, 2, 3, 4
	Discretization	noSteps	6, 9, 12
Section 3.3.1	Hysteresis	$hyst_{value}$	0.1 , 0.2 , 0.3
		$hyst_{maxtime}$	5, 10, 20, 50
Section 3.3.3	Symbolic Reduction	logBasis	NaN, e, 10
Section 3.4	Pattern Discovery	Algorithm	Adapted Sequitur, Pattern Enumeration

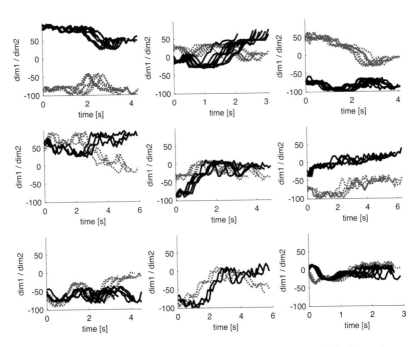

Figure 5.2 Exemplary visualization of patterns discovered in a 2D synthetic data set by pre-processing the time series with discretization, hysteresis, unification and logarithmic reduction

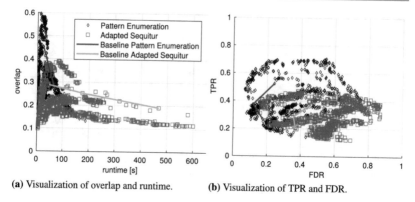

(a) Visualization of overlap and runtime.   **(b)** Visualization of TPR and FDR.

Figure 5.3 Results of parameter experiments with synthetic data set. Each parameter combination and its resulting overlap, runtime, TPR and FDR is shown in both figures by a marker. Squared markers show results of adapted Sequitur, while diamond markers show results of pattern enumeration. As a reference the results of the baseline method separately for each algorithm are visualized as solid lines[2]

Taking into account only the overlap, the best performing parameter setting is the one with the highest value in Fig. 5.3a. This value corresponds to a setting with $noSteps = 9$, no wavelet or PAA filtering, the usage of hysteresis with $hyst_{value} = 0.3$ and $hyst_{time} = 50$, standard symbolic reduction and the usage of pattern enumeration algorithm. In Figure 5.4 a more detailed analysis regarding the best matching patterns is shown for every predefined pattern. As previously described, an overlap is calculated for every discovered pattern combined with every predefined pattern. Then a best match for every predefined pattern is selected. These best matches regarding their symbolic length (gray line) as well as their overlap (blue markers) are shown in the figure. Beside this, the figure shows the symbolic length of every predefined pattern member preprocessed by the parameters that are mentioned above. Due to the fact, that the predefined patterns are too noisy to be compressed into an identical symbolic sequence, the symbolic length is not always the same for every member within one pattern. This is why the symbolic length of predefined patterns are visualized as boxplots with the red line marking the median length. The dark blue area shows the interval between the 25% and 75% quantile, while the

[2] "Intelligent Data Analysis, vol. 25, no. 5, Noering et al., Improving Discretization Based Pattern Discovery for Multivariate Time Series by Additional Preprocessing, pp. 1051–1072, (2021)", with permission from IOS Press. The publication is available at IOS Press through http://dx.doi.org/10.3233/IDA-205329.

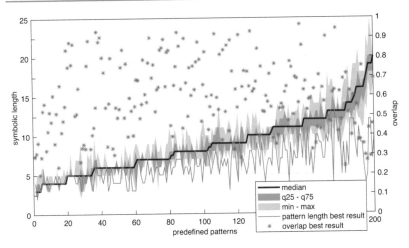

Figure 5.4 Symbolic length of predefined patterns after preprocessing the time series as well as the length and overlap of best pattern matches for every predefined pattern

light blue area shows the interval between the minimum and maximum symbolic length of a predefined pattern. A perfect match, and hence an overlap of one, would only be possible if minimum, maximum, 25% as well as 75% quantile values are all equal to the median. The major task of the preprocessing is to compress the time series, including all the predefined patterns, in a way that every member in a predefined pattern is assigned to the same symbolic sequence. This is possible for a subset of predefined patterns. But due to the high diversity that is designed into the synthetic patterns, it is definitely not possible for all patterns at once. This is why the overlap values of the predefined patterns have a high variance. While there are various patterns that are close to the perfect match, there are still patterns that only have an overlap of approximately 0.3. Additionally, the gray line in comparison to the boxplot shows, that the discovered patterns tend to cover only a part of the patterns, instead of covering considerably bigger patterns.

Figure 5.5 shows another detailed view on the results, highlighting the differences in quality and runtime between the filtering methods PAA, wavelet, hysteresis and no filtering at all. In general, any filtering method is better than no filtering, if it is well parametrized. The higher the number of bins is, the more important is filtering. Comparing wavelet and PAA filtering, Fig. 5.5 shows that in the best case PAA works better while also running faster. It also shows clearly that the results along the pareto front regarding the upper left corner are dominated by tests using

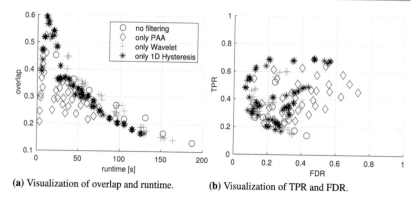

(a) Visualization of overlap and runtime.   **(b)** Visualization of TPR and FDR.

Figure 5.5 Comparison of filtering methods[3]

the proposed hysteresis approach. For a more detailed view see Fig. 5.6 with nine parameter settings being compared with fixed parameters except for the filtering parameters, which is meaningful for the comparison under equal circumstances. The fixed parameters $noSteps$ and $logBasis$ for each setting are written below the figure. A $logBasis$ of NaN refers to the standard reduction technique. Noticeable is the high variance of the overlap using PAA. Hence, a badly parametrized PAA can also produce results worse than without filtering at all. Focusing on the hysteresis the variance is less than in case of the PAA. Speaking of this synthetic test case, high values of the parameter $hyst_{value}$ and not too low values of $hyst_{maxtime}$ result in good quality. Note that both parameters need to be adapted to the signal to noise ratio in general. If the ratio is high, meaning there is only low noise, $hyst_{value}$ and $hyst_{maxtime}$ should be fixed to small values. In contrast, if the ratio is small, both parameters should have higher values. Additionally, the experiments have also shown that the hysteresis can also have a positive impact when using it on top of PAA especially with low values of $windowsize$.

To complete the analysis of this experiment, in Fig. 5.7 the results concerning the symbolic reduction are shown. The standard symbolic reduction, proposed in previous work, leads to good results concerning the overlap and the TPR. On the downside it also produces extremely high FDR values when the parameters are not set carefully, which is a consequence of the extreme reduction. By the use of the

---

[3] "Intelligent Data Analysis, vol. 25, no. 5, Noering et al., Improving Discretization Based Pattern Discovery for Multivariate Time Series by Additional Preprocessing, pp. 1051-1072, (2021)", with permission from IOS Press. The publication is available at IOS Press through http://dx.doi.org/10.3233/IDA-205329.

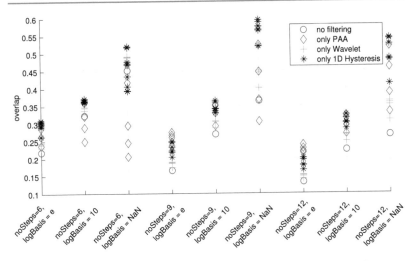

Figure 5.6 Overlap of different filter parameters, while fixing all other parameters. Parameter settings 1, 2, 4, 5, 7 and 8 used the logarithmic reduction. Parameter settings 3, 6 and 9 used the standard reduction technique[4]

logarithmic symbolic reduction it is possible to reduce FP and consequently the FDR in those cases. In this synthetic test case the higher accuracy also leads to a decrease in overlap and TPR, because the predefined patterns are designed to be time scaled. Hence, the same predefined pattern which can be perfectly detected by the standard symbolic reduction will be divided into multiple discovered patterns when using the logarithmic approach. To show the advantages of this effect in more detail, a real-life test is introduced in the next section.

5.1.2   Real-Life Data Set

Experiment Description
This experiment focuses on the evaluation of the logarithmic symbolic reduction in comparison to the standard symbolic reduction in terms of the trade-off between improving the quality and worsen the runtime. Therefore, a real-life data set of a

---

[4] "Intelligent Data Analysis, vol. 25, no. 5, Noering et al., Improving Discretization Based Pattern Discovery for Multivariate Time Series by Additional Preprocessing, pp. 1051-1072, (2021)", with permission from IOS Press. The publication is available at IOS Press through http://dx.doi.org/10.3233/IDA-205329.

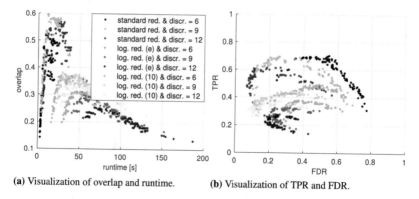

(a) Visualization of overlap and runtime. **(b)** Visualization of TPR and FDR.

Figure 5.7 Comparison of number of bins and symbolic reduction techniques[5]

battery electric vehicle is chosen with roughly 130 mio. samples and more than 2000 labeled charging events in the voltage signal. An extract of this data set is shown in Fig. 5.8 with charging events marked as solid lines. The goal of this experiment is to refind all the charging events by the use of the pattern discovery. Note that this use case is just a synthetic one in order to evaluate the performance of the method regarding a labeled data set. The actual task of finding charging events could be easily done without the pattern discovery. However, this is a hard task for the pattern discovery because the voltage within a charging event does not always behave in the same way. It depends on the State of Charge (SOC) at the beginning and in the end of a charging event, on the temperature, on the charging current and on some other effects. Hence, the start and end voltage, the gradient and the duration of a charging event strongly vary, which will be shown subsequently. Because of this variation it is not the goal to parametrize the pattern discovery in a way to classify all charging events as one pattern, but to find a set of patterns which describes the charging events as precisely as possible. Therefore, the precision also known as *Positive Predicted Value* (PPV) is additionally defined as:

$$PPV = \frac{|TP|}{|TP \cup FP|} = \frac{|TP|}{|I_{disc}|} = 1 - FDR \qquad (5.4)$$

[5] "Intelligent Data Analysis, vol. 25, no. 5, Noering et al., Improving Discretization Based Pattern Discovery for Multivariate Time Series by Additional Preprocessing, pp. 1051-1072, (2021)", with permission from IOS Press. The publication is available at IOS Press through http://dx.doi.org/10.3233/IDA-205329.

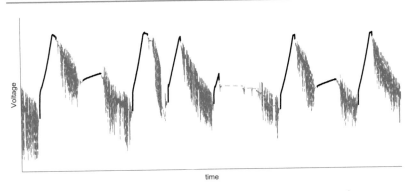

Figure 5.8 Voltage measurements with charging events marked as solid lines[6]

The same data set was tested with six different parameter settings varying the parameters $noSteps$ and $logBasis$ while fixing the other parameters regarding the results of the previous synthetic experiments. Hence, the hysteresis as the only filtering technique and the pattern enumeration algorithm was used. In Table 5.2 the results of this experiment are shown with an assignment concerning the parameters. For every parameter setting the number of patterns were evaluated, which are necessary to describe more than 90 percent of the charging events. For all those patterns the PPV and TPR were calculated. The average of the PPV and TPR for each parameter setting is shown in the table. Furthermore, by first building one global pattern per parameter setting from all those patterns that represent a subset of the charging events, the combined PPV and TPR were evaluated. Additionally, the runtime of all preprocessing functions and the pattern discovery algorithm overall is shown.

Results
As expected, the number of patterns to represent the charging events is increasing with the usage of the logarithmic symbolic reduction and its parameter $logBasis$. Hence, looking e.g. at the first column of Table 5.2, each pattern represents only a few charging events which also results in a low average TPR. To further visualize the variation of the charging events, Fig. 5.9a shows one member of every 31 patterns, derived from the parameters of column 2, in a different color. In Fig. 5.9b one exemplary pattern with its members and a representative is visualized. However,

---

[6] "Intelligent Data Analysis, vol. 25, no. 5, Noering et al., Improving Discretization Based Pattern Discovery for Multivariate Time Series by Additional Preprocessing, pp. 1051-1072, (2021)", with permission from IOS Press. The publication is available at IOS Press through http://dx.doi.org/10.3233/IDA-205329.

Table 5.2 Evaluation of real life experiment with six different parameter settings. The reduction with the value NaN means the usage of the standard symbolic reduction while the other values specify the parameter *logbasis* of the logarithmic symbolic reduction[7]

Reduction	e	10	NaN	e	10	NaN
Number of bins	6	6	6	9	9	9
Number of patterns	96	31	4	292	49	6
Average PPV	0.8723	0.8235	0.7239	0.8896	0.8611	0.7748
Average TPR	0.0320	0.0884	0.5328	0.0145	0.0760	0.4119
Combined PPV	0.7950	0.7498	0.5918	0.8034	0.7914	0.6300
Combined TPR	0.9427	0.9761	0.9530	0.9764	0.9617	0.9501
Runtime [min]	26.6	5.7	0.9	37.6	9.6	1.7

the average PPV as well as the combined PPV are much higher with the use of the logarithmic reduction in comparison to the standard reduction, meaning less false positives per pattern. At the same time every parameter setting is able to represent a similar amount of charging events, which is shown by the values of the combined TPR. The pattern discovery is now able to differentiate between a dynamic symbolic value change and a slow long-term value change by the use of the new proposed reduction technique. Because of the higher precision the logarithmic reduction technique can be seen as an additional built in clustering method. This of course comes with costs in runtime, because of the lower time series compression and the higher amount of discoverable patterns. Beyond that, the logarithmic reduction has a more positive impact when using a low number of bins.

However, what does that improvement mean to the RDC construction use case? Beside RDCs that are only based on the velocity of a vehicle, there may be additional requirements for the representativeness regarding the charging behavior of battery electric vehicles. Hence, an RDC may not contain the velocity, but other signals like the voltage or current of the battery. By applying the pattern discovery to the voltage signal, it is possible to extract a set of voltage patterns, that represents the driving as well as the charging of the vehicle. As we demonstrated previously, by means of the additional preprocessing methods, it is possible to differentiate between different kinds of charging events with less false positives than with previous approaches. The

---

[7] "Intelligent Data Analysis, vol. 25, no. 5, Noering et al., Improving Discretization Based Pattern Discovery for Multivariate Time Series by Additional Preprocessing, pp. 1051-1072, (2021)", with permission from IOS Press. The publication is available at IOS Press through http://dx.doi.org/10.3233/IDA-205329.

differentiation is necessary due to the high variance of charging events, as shown in Figure 5.9a, which causes e.g. different aging effects in the battery-cells. These variances should also be represented in a cycle.

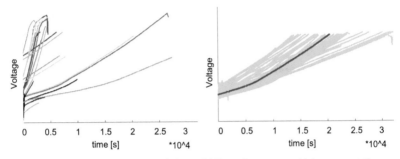

(a) Representatives of every rediscovered char-   (b) Exemplary pattern with its representative
ging pattern found by the pattern discovery   (highlighted).
algorithm ($logBasis = 10$, $noSteps = 6$).
Note that the colors repeat for different patterns due to insufficient availability of colors.

Figure 5.9 Differentiation of charging events by pattern discovery[8]

5.1.3   Dimensionality

This section shows a third experiment concerning the results in a growing dimensional space to show that this approach works in an environment with more than two dimensions. Therefore, a synthetic five-dimensional time series was generated by applying the random data generator from 5.1.1 with roughly 0.5 million samples per dimension. Afterwards, the pattern discovery was tested with a limited number of parameter settings and a varying number of dimensions, starting with two dimensions. The results are shown in Fig. 5.10. As expected, the quality of the results is slightly decreasing with increasing number of dimensions. This is because of the increasing complexity and the appearance of transitional states in the unified symbolic time series, which is sketched in Fig. 5.11 and Table 5.3. These two sequences are qualitatively similar to the human eye. But while going through the tool chain

[8] "Intelligent Data Analysis, vol. 25, no. 5, Noering et al., Improving Discretization Based Pattern Discovery for Multivariate Time Series by Additional Preprocessing, pp. 1051-1072, (2021)", with permission from IOS Press. The publication is available at IOS Press through http://dx.doi.org/10.3233/IDA-205329.

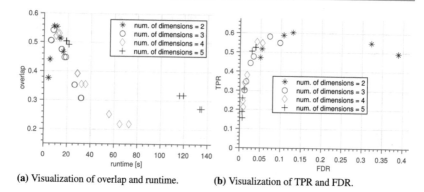

(a) Visualization of overlap and runtime.   **(b)** Visualization of TPR and FDR.

Figure 5.10 Test of the pattern discovery with varying number of dimensions[9]

with both of these sequences, it is evident that they have not the same sequence of symbols after the unification process, which is shown in the "unified" rows of Table 5.3. Even though they both go from symbol 2 to symbol 3, they have different transitional states, which makes it impossible for the pattern discovery to detect both sequences as the same pattern. This problem gets worse the more dimensions are included and, hence, is a bottleneck of the proposed unification approach. Nevertheless, Fig. 5.10 shows that the pattern discovery tool chain achieves acceptable results even for five-dimensional analyses. A strategy for good results in high-dimensional spaces is the reduction of the number of bins per dimension.

5.1.4   Postprocessing

In previous Sections 5.1.1 to 5.1.3 the pattern discovery without postprocessing was tested. This section evaluates if the postprocessing techniques proposed in Section 3.5 are able to reduce the amount of patterns while maintaining the relevance of discovered patterns. Therefore, the best performing parameter setting from Section 5.1.1 is chosen and again applied to the synthetic time series. Based on these results multiple postprocessing settings based on weighted greedy approach are evaluated concerning their overlap and number of patterns. Additionally, the repre-

---

[9] "Intelligent Data Analysis, vol. 25, no. 5, Noering et al., Improving Discretization Based Pattern Discovery for Multivariate Time Series by Additional Preprocessing, pp. 1051-1072, (2021)", with permission from IOS Press. The publication is available at IOS Press through http://dx.doi.org/10.3233/IDA-205329.

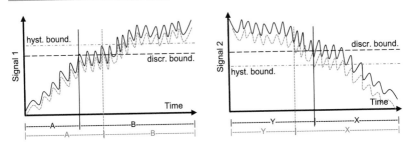

Figure 5.11 Two similar multivariate sequences being discretized. The unification process is shown in Table 5.3[10]

Table 5.3 Unification of two multivariate sequences from Fig. 5.11 using the same translation table as in Fig. 3.4. The first multivariate sequence starts by changing its discretized value from A to B within the space of signal 1. Subsequently, it also changes its value from Y to X within the space of signal 2. Looking at sequence 2, this order is switched. At first there is a change in signal 2 and afterwards in signal 1. Due to the switched order the transitional state of these two sequences vary. The transitionales states are written in bold type

1st seq.	Signal 1	A	B	B
	Signal 2	Y	Y	X
	Unified	2	**4**	3
2nd seq.	Signal 1	A	A	B
	Signal 2	Y	X	X
	Unified	2	**1**	3

sentativeness is analyzed by having a look on each index of the time series and how often it is represented by a pattern.

The best performing parameters include a discretization with $noSteps = 9$, no wavelet or PAA filtering, the usage of hysteresis with $hyst_{value} = 0.3$ and $hyst_{time} = 50$, standard symbolic reduction and the usage of pattern enumeration algorithm. The parameter space of the postprocessing, shown in Table 5.4, is searched by brute force. Beside the selection of a weighting method and the α-threshold for the greedy algorithm, the minimum acceptable size of a pattern is

---

[10] "Intelligent Data Analysis, vol. 25, no. 5, Noering et al., Improving Discretization Based Pattern Discovery for Multivariate Time Series by Additional Preprocessing, pp. 1051-1072, (2021)", with permission from IOS Press. The publication is available at IOS Press through http://dx.doi.org/10.3233/IDA-205329.

Table 5.4 Parameter space of experiment for postprocessing

Parameter	Parameter values
Weights	standard, static, dynamic, skyline
α	0.8 , 0.9 , 0.95 , 0.99
MinPatternSize	2, 3, 4, 5

varied. This parameter is added to the evaluation, because by experience there is a sensitivity of the results when ignoring patterns with small symbolic length.

Relevance of Patterns

In Figure 5.12 the results of the brute force search is shown regarding the overlap, number of patterns as well as the relation between them. On the x-axis the parameters α and $MinPatternSize$ are varied, while the colored lines correspond to the variants of the greedy postprocessing approach. Additionally, reference results without postprocessing and the usage of the pattern enumeration algorithm as well as the Sequitur algorithm are marked as dashed lines.

Every greedy variant is able to select a set of patterns that covers the predefined patterns nearly as good as the pattern discovery without postprocessing (0.59 to > 0.52), while at the same time reducing the amount of patterns in the worst case to approximately one tenth (from 67,733 to 6,293). If an imaginary threshold is set to an overlap of 0.5 the number of patterns can be reduced to approximately 1% of the initial pattern number (from 67,733 to 790) by the usage of static greedy algorithm. One may have expected a number of patterns of around 200, because the synthetic data set is spiked with 200 different predefined patterns.

Note that these 200 patterns only cover approximately 44% of the time series, the rest is random. In contrast to this, the minimum coverage (α) is set to at least 80%. Hence, many discovered patterns represent random parts of the time series.

Due to the built in reduction technique of the Sequitur algorithm, it is necessary to additionally compare the greedy postprocessing approaches regarding the Sequitur. As shown in the figure the Sequitur algorithm is able to discover patterns with an overlap of approximately 0.38 while reducing the amount of patterns to 3,547. As shown in the figure, every greedy approach is able to select a pattern set with better overlap and less patterns than the Sequitur algorithm. For example using the static greedy approach with a $MinPatternSize$ of 3 and an α threshold of 0.8 results in an overlap similar to the usage of Sequitur, but requiring only one tenth of patterns (3,547 to 361).

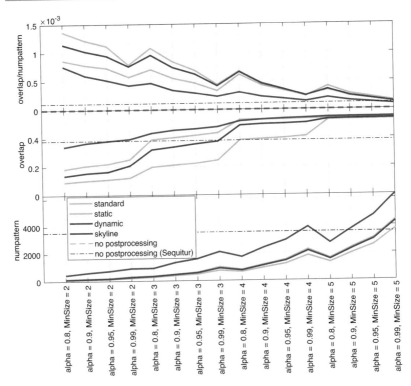

Figure 5.12 Overlap and number of patterns discovered by weighted greedy postprocessing technique with varying parameter settings. Comparison to results without postprocessing

The standard greedy approach can be highly improved by including other figures in the selection process like the (unique) symbolic length and the number of patterns. Especially the variants static and dynamic have a positive effect on the overlap while not increasing the number of patterns to be required. As already described earlier, the skyline greedy approach leads to a colorful selection of patterns. Hence, the task of finding the smallest set of patterns to cover the time series is not strictly followed when using the skyline approach. This is why it consequently leads to a higher number of patterns. On the other side, the high amount of patterns increases the probability to cover all predefined patterns, which leads to high overlap values.

Apparently, the standard greedy approach leads to unacceptably low overlap values with a *MinPatternSize* of 2. This is because it has a problem with the high diversity and randomness of the synthetic time series (56 % is random). In Figure 5.13a the relative distribution of discovered pattern lengths is shown for every greedy approach and additionally for the predefined patterns. When taking into account all possible pattern lengths (2 to infinity), the standard greedy algorithm selects only patterns with a length of two, which definitely is a failure. This problem can be weakened by switching to dynamic, static or skyline greedy algorithm. Beside this, a slight increase of the minimum pattern size has a positive effect on the distribution of pattern lengths, as shown in Figure 5.13b with the minimum pattern length of 5. As already described, this also leads to highly improved overlap values, especially when using standard, static or dynamic greedy approach.

In general, the value of α positively correlates with the overlap independently of the chosen greedy approach. But in the range of 0.8 and 0.99 an increase of α does not lead to a high boost of overlap. While on the other hand the number of patterns increases much stronger. This probably is, because the majority of the time series is random and with a pattern set that already covers 80 % of the time series, the probability is already quite high to cover the most of the predefined patterns.

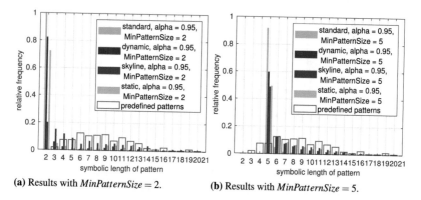

(a) Results with *MinPatternSize* = 2.   (b) Results with *MinPatternSize* = 5.

Figure 5.13 Distribution of symbolic pattern length for every discovered pattern and different postprocessing parameter settings in comparison to the distribution of predefined patterns

Representativeness of Patterns

In the previous section it was shown that the pattern discovery is able to identify relevant patterns. Now what is about the representativeness? The most representative set of patterns would cover every index of the time series exactly once. The highest goal of the standard greedy approach is to identify the smallest set of patterns that best covers the majority of the time series. In order to additionally satisfy the constraint of

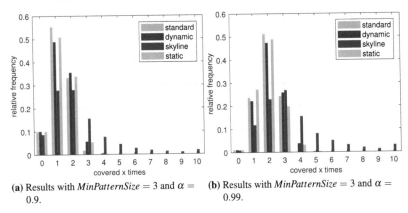

(a) Results with *MinPatternSize* = 3 and α = 0.9.

(b) Results with *MinPatternSize* = 3 and α = 0.99.

Figure 5.14 Proportion of the time series covered 0 to 10 times by applying all four greedy approaches

representativeness, the variable $overlap2Represented$ was implemented into every proposed greedy approach. Figure 5.14a shows the proportion of time series indices that is covered x times for all four greedy approaches and a fixed parameter setting. In *standard*, *dynamic* and *static* mode the postprocessing is able to discover patterns that cover approximately 50 % of the time series exactly once. The remaining 40 % are majorly divided to being covered two and three times. The skyline greedy approach is again an exception. It shows a tendency of covering the major part of the time series multiple times, because the task of finding the smallest set of patterns is not strictly followed.

While Fig. 5.14a shows the multiple coverage with an α value of 90 %, in Fig. 5.14b α was increased to 99 %. In general, a shift to the right can be observed. In *standard*, *dynamic* and *static* mode now approximately 50 % of the time series is covered twice instead of once. Only slightly more than 20 % is covered exactly once. This is a common phenomenon when increasing the α value. Depending on the data set, the beginning of majorly multiple coverage can vary. It usually starts when all frequently occurring patterns are discovered and the rest of the uncovered time series is random or unique. This phenomenon could theoretically be utilized to implement a dynamic α value, which adapts to the data set. For example the postprocessing would stop when the amount of multiply covered indices exceed the amount of indices that are covered exactly once.

5.2 Representativeness of Driving Cycles

This section aims to evaluate the effectiveness of proposed pattern-based RDC construction approaches. Is it possible to construct or identify RDCs by pattern-based approaches without even inputting information regarding the representativeness? Are these unsupervised approaches able to compete with knowledge-based approaches, that do have information regarding the characteristic parameters? In order to answer these questions, Section 5.2.1 introduces three evaluation criteria. While the first one is used to identify a knowledge-based RDC as a reference for pattern-based approaches, the second and third one will be used as independent evaluation criteria. In Section 5.2.2 the pattern-based full trip RDC approach will be evaluated against the knowledge-based approach. In Section 5.2.3 follows the equivalent evaluation of the pattern-based RDC construction technique. In Sections 5.2.2 and 5.2.3 the approaches will be tested with respect to the use cases 2 and 3 described in Section 4.1.2. Hence, the objective is to find an RDC that is representative regarding the frequency of patterns. In contrast to this, Section 5.2.4 will deal with the construction of RDCs under consideration of use case 1, meaning the shortest possible cycle that contains every pattern just once.

The evaluation will be primarily based on one fixed parameter setting for the pattern discovery. The settings and exemplary results of the pattern discovery are shown in Section 5.2.1. There might be improvements regarding the representativeness of pattern-based RDCs with tuned pattern discovery parameter settings. Nevertheless, this section shows that it is not necessary to fine-tune the pattern discovery settings in order to get good results.

5.2.1 Basics

Before going into detail concerning the evaluation of both RDC construction techniques, that were proposed in Chapter 4, this section describes three evaluation criteria and gives a brief overview of the pattern discovery parameters and results that will be used as a basis for the pattern-based RDC construction. The first evaluation criterion will be used to select a knowledge-based RDC as a reference for pattern-based approaches in later sections. The second one will be used as an independent quality measure in order to evaluate the representativeness of knowledge-based approaches regarding features that were not directly included into their feature space.

The data set under consideration contains real driving data of eight vehicles of the Volkswagen Group used in a taxi-similar context. In total, they drove approximately 172,000 kilometers on public streets and logged data in a sampling rate of

20 Hz. This results in approximately 567 mio. samples that are cut into 2700 trips with varying length by the method that was described in Section 4.2.1. Within the evaluation of pattern-based RDCs in Sections 5.2.2 to 5.2.4 only the velocity signal is used. This is because of comparability between both approaches. While the full trip identification approach could easily handle multivariate time series, the RDC construction technique is only made to construct univariate cycles.

Evaluation Criterion 1
To evaluate the performance of the proposed pattern-based techniques, a competing knowledge-based full trip approach is implemented. In Section 2.2.2 possible knowledge-based features ere already listed. For evaluation, the most commonly used features are used to identify an RDC from a set of trips. These features are shown in Table 5.5 and calculated for every trip separately. The table captures the minimum, maximum, median, mean and standard deviation of all features for all the trips. Additionally, it shows the features of an exemplary RDC and the relative error of the RDCs features regarding their median values.
First of all, the VSP, as an equivalent to the energy consumption, is calculated for every trip by calculating the driving resistances for the trips velocity trajectory as shown in the following equations. For simplicity, the calculations are based on the assumption of absence of slope of the street. Hence, the driving resistances include the aerodynamic drag force F_{AD}, the rolling resistance force F_{RR} and the acceleration force F_{Acc}:

$$F_{AD} = c_w \times A \times \frac{\rho_{Air}}{2} \times v^2 \tag{5.5}$$

$$F_{RR} = m_v \times g \times f_{RR} \tag{5.6}$$

$$F_{Acc} = m_v \times a \tag{5.7}$$

Note that, while the velocity v is directly derived from the logged data, the longitudinal acceleration a needs to be recalculated based on the velocity in order to avoid the influence of the slope of the street. For the air density ρ_{Air}, the rolling resistance coefficient f_{RR}, as well as for the gravity constant g standard values were assumed. The other parameters are specifically determined for the vehicles:

- Vehicle mass: $m_v = 3500 kg$
- Frontal area: $A = 4.7 m^2$
- Aerodynamic drag coefficient: $c_w = 0.33$

Based on these forces the total power P to move the vehicle, as well as the VSP can be calculated as:

$$P = (F_{AD} + F_{RR} + F_{Acc}) \times v \tag{5.8}$$

$$VSP = \frac{\int P(P > 0)dt}{distance} \tag{5.9}$$

Note that the VSP does not describe the real energy consumption, due to the fact that no powertrain specific parameters were used for this simulation. It just reflects the amount of energy that is necessary for propulsion of the vehicle. This is why only the positive power values are used for the calculation of the VSP.

Beside the VSP, some commonly used statistic parameters like mean, median and standard deviation of velocity, acceleration and deceleration are included as features for the knowledge-based RDC. Every sample with a longitudinal acceleration greater than zero is considered as acceleration, while every sample smaller than zero corresponds to a deceleration. Note that the standard deviation is only calculated for the longitudinal acceleration without differentiation. Furthermore, the time spend in states like cruising, acceleration, deceleration and idling as proposed in [14] is considered in the knowledge-based RDC. They partitioned the data into these states based on four parameters (a_1, a_2, δ, n). Additionally, they determined optimal parameters in this study, which is why this work assumes the same parameter values.

After calculating these features for all 2700 trips, PCA is applied to avoid biased results regarding overrepresented features. In order to explain 99.9 % of the original feature information, five new features result from the PCA. Within this five-dimensional space the Euclidean distance between every pair of trips is calculated to identify the most representative trip.

For further evaluation, a feature-based quality value ($quality_{feat}$) for every trip is calculated by averaging the relative error column in Table 5.5.

Evaluation Criterion 2

To highlight the advantages of pattern-based RDC construction techniques in contrast to knowledge-based approaches a second evaluation criterion is introduced. Beside the features defined in the previous section, it may be of high interest that the RDC covers all the velocity and longitudinal acceleration states that occur in real-life driving. In Figures 5.15a and 5.15b a relative frequency histogram of both the velocity and acceleration based on the input data set is visualized as blue bars. In the best case the histogram based on the RDC would show exactly the same distribution. In both figures an exemplary RDC is additionally plotted as orange bars. Going one step further, Fig. 5.15c shows the two-dimensional histogram based

Table 5.5 Statistics for the data set as well as for an exemplary RDC constructed by a pattern-based approach

	Min.	Max.	Median	Mean	Std.	RDC	Relative error [%]
VSP	0	66.68	28.22	27.97	4.18	30.93	9.6
Mean acceleration	0	5.70	2.22	2.20	0.33	2.40	8.4
Median acceleration	0	4.57	1.50	1.47	0.25	1.87	21.4
Mean deceleration	−3.85	0	−2.07	−2.07	0.28	−2.30	9.9
Median deceleration	−2.80	0	−1.34	−1.33	0.19	−1.67	20
Std. long. acceleration	0	20.70	2.55	2.35	0.94	2.06	12.4
Mean velocity	0	95.08	20.95	21.02	5.86	17.49	16.8
Std. velocity	0	57.29	21.47	21.02	4.27	18.79	10.6
Time in acceleration	0	0.28	0.10	0.10	0.03	0.12	2.4
Time in deceleration	0	0.27	0.09	0.10	0.03	0.12	2.4
Time in cruising	0	0.81	0.43	0.44	0.10	0.34	10.1
Time in idling	0.09	1	0.39	0.36	0.13	0.43	6.9

on the real-life input data set as an image with high relative frequencies colored in yellow and low relative frequencies colored in dark blue. Likewise, Fig. 5.15d shows the image based on an exemplary RDC. Both two-dimensional histograms can be used to gain a quality measure for the RDC by calculating the absolute differences of both and summing up the delta values:

$$quality_{hist} = 1 - \sum_{i,j} \left| hist2D_{orig}(i, j) - hist2D_{RDC}(i, j) \right| \qquad (5.10)$$

Evaluation Criterion 3
Especially for the evaluation of RDCs constructed to fit use case 1 another evaluation criterion has to be applied. Previous criteria defined the representativeness of an RDC as knowledge-based features whose values are as close as possible to the feature values of the underlying data set. In use case 1 these features are not important anymore, because the goal is to construct the shortest possible cycle that includes all the existing situations independently of their frequency. Hence, features like the VSP or average speed do not express the quality of a cycle. This is why another criterion is introduced, which is based on the idea of criterion 2. The basic idea is, that an RDC

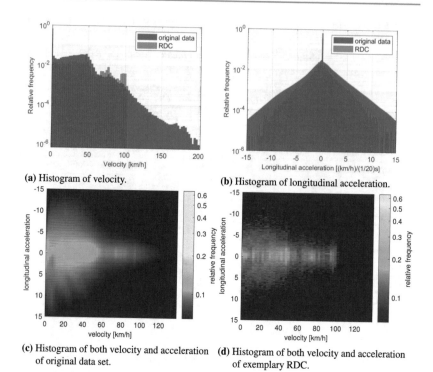

(a) Histogram of velocity.

(b) Histogram of longitudinal acceleration.

(c) Histogram of both velocity and acceleration of original data set.

(d) Histogram of both velocity and acceleration of exemplary RDC.

Figure 5.15 Evaluation criterion 2

needs to run through every state defined by velocity and longitudinal acceleration. In Figure 5.16 this concept is visualized. In equivalence to the previous criterion, this figure shows the binary histogram of both signals. In Fig. 5.16a the original data is shown, while Fig. 5.16b visualizes the RDC. Every state in the histogram is marked as white dot, if it occurred at least once in the corresponding data set. The binary quality measure is determined by calculating the absolute difference of both matrices in relation to the total number of states:

$$quality_{binaryhist} = 1 - \frac{\sum_{i,j} \left| binaryhist2D_{orig}(i,j) - binaryhist2D_{RDC}(i,j) \right|}{numOfStates}$$

(5.11)

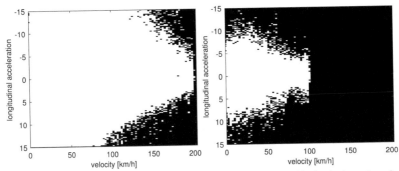

(a) Binary histogram of both velocity and accel-   **(b)** Binary histogram of both velocity and acceleration of original data set.   eration of exemplary RDC.

Figure 5.16 Evaluation criterion 3 for evently distributed pattern frequency

Pattern Discovery for RDC Construction

To limit the effort of evaluation for the pattern-based RDC approaches, the pattern discovery settings are fixed beforehand. The selected settings are listed in Table 5.6. They are based on the experience gained from the pattern discovery evaluation in Section 5.1. The time series was discretized into 40 equal distant bins, which results in a bin size of approximately 5 km/h. Additionally, another discretization boundary was included at 2 km/h in order to enhance the differentiation between standstill and driving of the vehicle. The filtering is realized by the proposed hysteresis technique. As both evaluation criteria explicitly include the acceleration of the vehicle, the logarithmic symbolic reduction needs to be chosen here. Otherwise the pattern discovery does not take into account the duration of situations. Hence, information about the acceleration would be ignored in the stage of pattern discovery. Beside this, the pattern enumeration algorithm is chosen as basis for the postprocessing, which is parametrized to select a pattern set that at least covers 90 % of the time series.

The pattern discovery tool chain discovered 255 patterns. The 16 most covering patterns are shown in Figure 5.17. Additionally, the APF and RPC values of all patterns are shown in Fig. 5.18a. In equivalence to Section 5.1.4 Figure 5.18b visualizes the proportion of the time series that is represented x times. In general, it shows the same phenomenon as mentioned earlier, which describes the shift to multiply covered indices. Reducing the α value may solve this special issue of representativeness. On the other hand it is necessary for the pattern-based RDC approach to discover a set of patterns that covers the highest possible portion of the time series.

Table 5.6 Parameter space of pattern discovery for RDC evaluation

Method	Parameter	Parameter values
Discretization	noSteps	40
Hysteresis	$hyst_{value}$	0.1
	$hyst_{maxtime}$	10
Symbolic Reduction	logBasis	10
Pattern Discovery	Algorithm	Pattern Enumeration
Postprocessing	Weights	dynamic
	α	0.9
	MinPatternSize	3

Fig. 5.18b additionally shows the proportions, when the dictionary is corrected by the approach described in Section 4.3.1. The amount of indices covered 4 or more times can be reduced by this method, which is important for the construction of pattern-based RDCs treated in Sections 5.2.3 and 5.2.4.

5.2.2   Identification of Full Trip RDC

This section deals with the evaluation of full trip RDCs based on the discovery of patterns proposed in Section 4.2 in comparison to a knowledge-based approach. Because the pattern-based approach heavily relies on the distance metric that is used to find the RDC, this section starts with an evaluation of different distance metrics. Subsequently, the results based on the best distance metrics are evaluated against the knowledge-based RDC approach under consideration of both quality measures.

Meaningfulness of Distance Measures
A major challenge of the pattern-based full trip RDC approach is the curse of dimensionality. By applying pattern discovery, the problem of selecting an RDC is translated into a high-dimensional space. Within this space a trip is described as APF or RPC of every discovered pattern. In order to solve this high-dimensional nearest neighbor problem, a distance metric is necessary that produces reliable results. As already described in Section 4.1.3, the authors of [2] proposed the usage of fractional L_k distance metrics with $k < 1$ in contrast to Euclidean distance ($k = 2$). This section examines the meaningfulness of distance metrics in the context of RDCs by varying the parameter k, the dimensionality, the input information (APF or RPC) and

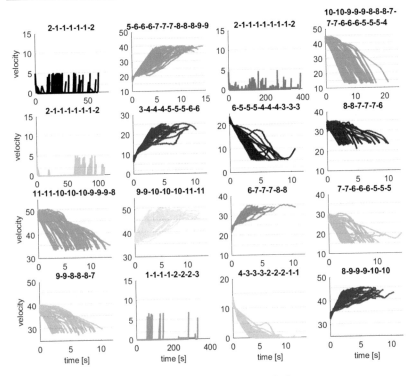

Figure 5.17 16 most covering patterns discovered in real-life data set

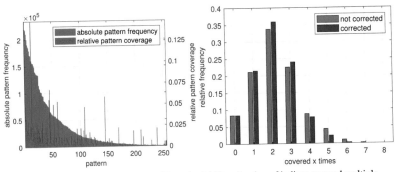

(a) RPC and APF of patterns discovered in real-life data set.

(b) Visualization of indices covered multiple times.

Figure 5.18 Pattern statistics of real-life data set

the normalization. The tested parameter values are shown in Table 5.7. For every parameter setting the relative contrast is calculated as defined in Equation (4.4), which is a measure for the meaningfulness of a distance metric.

Table 5.7 Parameter space for curse of dimensionality test

Parameter	Parameter values
Input	APF, RPC
Normalization	true, false
k	0.25, 0.5, 0.75, 1, 2
Dimensions	1, 29, 57, 86, 114, 142, 170, 199, 227, 255

In [2] the authors made a statement, that the meaningfulness of fractional distance metrics worsens slower with increasing dimensionality in comparison to distance metrics with $k \geq 1$. In order to check this statement in the context of RDCs, in Figure 5.19 the development of relative contrast with increasing dimensionality is shown for all the parameter settings. To make a statement regarding which distance metric worsens the slowest, all of these relative contrast graphs were referenced to their maximum. In general, all graphs show the expected decrease of meaningfulness with increasing dimensionality. But in contrast to the statements from the authors of [2], in three of four cases the meaningfulness of the Euclidean distance decreases the least instead of the most. This observation goes along with findings from Klawonn et al. in [48]. Amongst other things, the authors studied the *concentration of norm phenomenon* (CoN), which says that for increasing dimensions the relative difference of the closest and the farthest point from the origin in the data set converges to zero. They demonstrated this phenomenon by means of a synthetic clustering use case with increasing dimensionality. In this use case it was impossible to seperate two simple clusters at a dimensionality of above 30 dimensions. Beyond this, they were able to show that the CoN phenomenon can be weakened by increasing k of the distance metric.

Hence, in case of the usage of non-normalized APF-spectra (Fig. 5.19a), non-normalized RPC-spectra (Fig. 5.19c) and normalized RPC-spectra (Fig. 5.19d) it is recommended to use the Euclidean over fractional distance metrics. When using the normalized APF-spectra there is only small variance when switching between the k values. Beside this, it is noticeable that the meaningfulness, especially with $k = 0.25$, again increases when exceeding 140 dimensions. At this point it is necessary to note, that this is only a snapshot of a curse of dimensionality analysis. These results also depend on the underlying data. Hence, effects like the sudden increase of meaningfulness in Fig. 5.19b or the toggling in Fig. 5.19d can also be caused by special features in the APF- or RPC-spectra.

Additionally, Figure 5.20 shows the non-normalized relative contrast when using 255 dimensions and varying k. Interestingly, the graphs for APF and RPC show opposite behavior. While the meaningfulness of RPC-based distance metrics increase with increasing k, the APF-based metrics show a negative tendency. When only considering this analysis, the suggestion would be to choose a fractional distance metric with $k = 0.25$ when using the APF as a basis and the Euclidean distance when using the RPC. Nevertheless, due to the more stable behavior with varying dimensions as shown in Figure 5.19 it is recommended to use the Euclidean distance over fractional distance metrics in the context of RDC. The findings of [48] reinforce this recommendation.

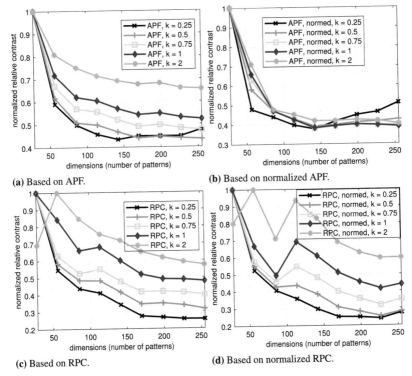

(a) Based on APF.

(b) Based on normalized APF.

(c) Based on RPC.

(d) Based on normalized RPC.

Figure 5.19 Normalized relative contrast for different distance metrics and varying dimensions

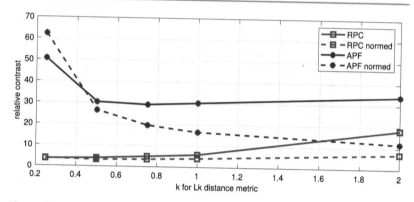

Figure 5.20 Relative contrast with varying k

Pattern-Based vs. Knowledge-Based

After working out the best choice in terms of distance metric, the evaluation of which input to use in order to achieve the best results regarding both knowledge-based criteria can be done. The results based on the Euclidean distance are visualized in Figure 5.21 for the four different types of input, as well as for the knowledge-based reference. It shows the cumulative histograms of both quality measures based on all 2700 trips. As expected, the knowledge-based RDC achieves the highest representativeness in terms of $quality_{feat}$. That makes sense because the same features were taken into account when selecting the RDC and calculating the quality measure. Nevertheless, it was not able to reach the highest value for $quality_{feat}$, due to the usage of PCA before the selection process. The knowledge-based RDC also achieves a high value in terms of $quality_{hist}$, due to the fact that the underlying features strongly correlate.

In general, the pattern-based RDCs are able to achieve acceptable results regarding both quality measures. In any input case the quality is higher than with random guessing, that would result in an average $quality_{feat}$ of approximately 0.91. Considering the second evaluation criterion, the benefit regarding random guessing is even bigger. Exceeding the random guessing approach is the minimum requirement. Beyond that, using both types of inputs APF and RPC the results can be improved by the proposed normalization technique, while the benefit when using the RPC with normalization seems to be even bigger. This leads to a $quality_{hist}$ that even exceeds the quality of knowledge-based methods.

To verify this statement the same analysis has been performed for two more pattern discovery settings. Within the second test case α was raised from 0.9 to 0.95 in

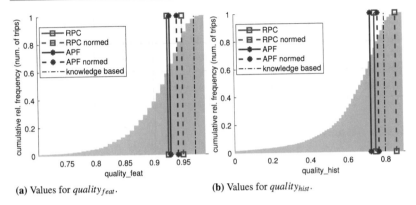

(a) Values for *quality*$_{feat}$.

(b) Values for *quality*$_{hist}$.

Figure 5.21 Cumulative relative frequency of quality values for all trips including different results from pattern-based full trip, as well as from the knowledge-based approach

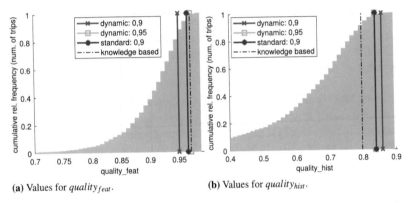

(a) Values for *quality*$_{feat}$.

(b) Values for *quality*$_{hist}$.

Figure 5.22 Cumulative relative frequency of quality values for all trips including results from pattern-based full trip based on three different settings of pattern discovery. In all three cases the Euclidean distance in combination with the normed RPC has been used

order to discover pattern that cover a greater part of the data set. The third test case switched the *weight* of the greedy algorithm in the postprocessing from dynamic to standard. All the other parameters stayed the same. The tendency of previous paragraphs proved to be true. The best results can be achieved by applying the Euclidean distance based on the normed RPC. In Figure 5.22 all three results are shown for *quality*$_{feat}$ and *quality*$_{hist}$. The representativeness in terms of *quality*$_{feat}$ was improved by slightly varying the parameters. In the second test case, the pattern-based full trip approach was even able to reach the same quality as the knowledge-based approach. In terms of $quality_{hist}$ all three parameter settings outperform the knowledge-based approach. To conclude this, the pattern-based RDC full trip identification has the potential of selecting the best balanced RDC without even inputting information regarding the characteristic parameters.

5.2.3   Construction of RDC

After showing the performance of pattern-based RDC identification techniques in the previous section, now the pattern-based RDC construction is evaluated regarding the use case 2 of Section 4.1.2. In general, the construction of cycles is much more complex than the identification. This is why this section begins with the description of challenges in practice resulting from the selection of a representative pattern set. Then, the constructed RDC is validated regarding the physical correctness, followed by the analysis of pattern frequency and coverage in different RDCs. Subsequently, the constructed RDC is evaluated against the knowledge-based approach, as well as the pattern-based RDC identification technique regarding the proposed quality measures.

Trade-off between Representativeness and Computing Time
In order to select a representative set of patterns for the construction of RDCs Section 4.1.2 introduced the possibility to reduce the absolute frequency of every pattern by the count of the rarest pattern. Unfortunately, in practice this procedure causes the representative pattern set to show an extremely high number of patterns. In Figure 5.23 the number of patterns and the resulting error of the RDCs RPC-spectrum regarding the original RPC-spectrum is visualized for different norm factors. The solid line shows the error with every pattern occurring at least once in the RDC, while the dashed line states the error with zero occurrences allowed. As shown by the diamond marker, the normalization by the count of the rarest pattern causes the representative pattern set to contain 48,570 patterns. Due to the high complexity of the ATSP, the computation of a solution for an ATSP with 255 nodes already needed approximately 2 days. This is why the user has to make a compromise between representativeness and computing time. In this case an exemplary trade-off was chosen, with approximately 500 patterns in total.

Physical Correctness
One challenge when synthetically constructing cycles is physical correctness. The synthetic cycle must not exceed the boundaries that are given by the vehicle-

physics. One possibility to verify this correctness is to check for implausible speed-acceleration states. Therefore the calculations of evaluation criterion 3 is utilized. In theory, a state that does not occur in the original data but in the RDC is an implausible state. Hence, every negative value in the following matrix $binarydelta$ is not allowed:

$$binarydelta = binaryhist2D_{orig} - binaryhist2D_{RDC} \qquad (5.12)$$

This matrix is visualized in Figure 5.24. If there was a negative value, it would be marked. But there are none. This is a first proof for the RDC to be physically correct. An excerpt of the tested RDC is shown in Figure 5.25. While Fig. 5.25a visualizes the concatenated patterns, Fig. 5.25b displays the resulting trip including overlap and smoothing of transitions.

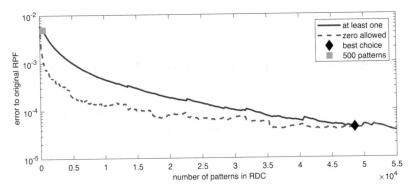

Figure 5.23 Number of patterns included in constructed RDC vs. error regarding the RPC of original data

Patterns in RDC
The representative pattern set, that was chosen as a trade-off between representativeness and computing time, is visualized in Figure 5.26 by framed bars. In theory, every of those patterns should be rediscovered in the RDC. Hence, when preprocessing the RDCs time series we should be able to rediscover the symbolic sequences of each pattern in the frequency that was chosen before. In the figure the rediscovered patterns frequencies are shown as filled bars. It is evident that these are not equal. Especially when concatenating patterns with symbolic overlap the symbolic sequences are not guaranteed to remain exactly the same, even though the pat-

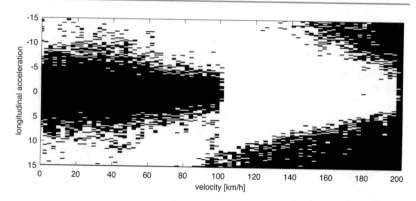

Figure 5.24 Visualization of *binarydelta*. Each white state is part of the original data, but not part of the RDC. Each black state is part of either both data sets or no data set at all. States that occur in the RDC but not in the original data would be highlighted, but there are none

terns had these symbolic sequences, which has three reasons. Firstly, the hysteresis filtering blurs the transitions by changing the symbolic sequences as described in Section 3.3.1. For example in the case of two patterns *AB* and *BA* being concatenated and the last value of *AB* as well as the first value of *BA* are within the hysteresis region of *B*, this may cause the preprocessing of the RDC to skip the value *B*. Secondly, the logarithmic symbolic reduction has similar effects to the RDCs symbolic representation. If two symbolic sequences being concatenated have the same connecting symbol, the logarithmic reduction might cause this symbol to be logarithmically reduced in another way than in the pattern itself. Thirdly, to construct a physically correct RDC there may be the need to fill gaps with patterns. In general, this procedure also causes the rediscovered APF-spectrum to vary. But in this explicit RDC no gap had to be filled with patterns.

Additionally, Figure 5.27a shows the RPC-spectra of the original data, the pattern-based constructed and identified RDC, as well as from the knowledge-based RDC. Due to the extreme compromise regarding representativeness and computing time, that had to be taken, the RPC-spectrum of the constructed RDC visually differs the most from the original data. To clarify this, Fig. 5.27b shows the theoretical cumulative RPC of each RDC if there was no overlap of any pattern. While the identified RDC as well as the knowledge-based RDC comes close to the original data, the constructed RDC shows a greater deviation, which is caused by the trade-off. If no compromise had to be taken, the black cumulative RPC would result. This analysis clearly shows the influence of the trade-off regarding the achievable representativeness (Figure 5.27).

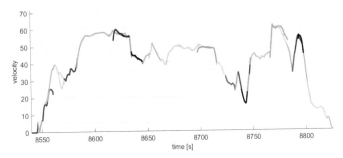

(a) Concatenated patterns.

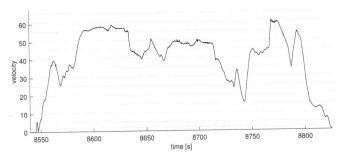

(b) Resulting trip including overlap and smoothing.

Figure 5.25 Excerpt of a constructed RDC

Pattern-Based vs. Knowledge-Based
In Figure 5.28 the quality of the constructed RDC in comparison to the identified as well as the knowledge-based RDC is shown. Unfortunately, the constructed RDC can not compete with the other approaches. This is definitely caused by the trade-off to the drawback of representativeness. If there was a faster algorithm to find a solution for the TSP, it would be possible to finally make a statement regarding the representativeness of the best possible constructed RDC. Until then, the amount of patterns to concatenate needs to be limited in order to restrict the computing time. Within the tested RDC 537 patterns were concatenated. To find a good solution the algorithm to solve the TSP needed approximately 12 hours. As outlined before, the representativeness of the constructed RDC is majorly influenced by the trade-off, which is why an evaluation regarding the benefit of overlapping patterns in contrast to non-overlapping patterns is unfeasible.

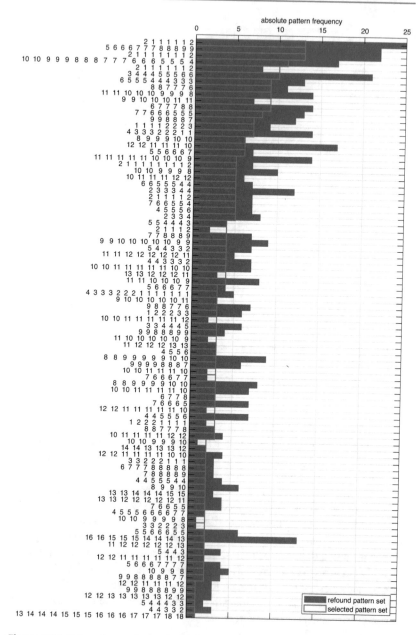

Figure 5.26 Selected vs. rediscovered representative pattern set (excerpt of 90 patterns)

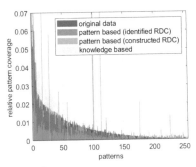 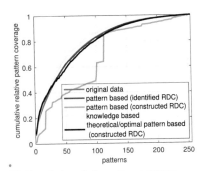

(a) Relative pattern coverage. **(b)** Cumulative relative pattern coverage.

Figure 5.27 Patterns and their coverage in the original data as well as in three different RDCs

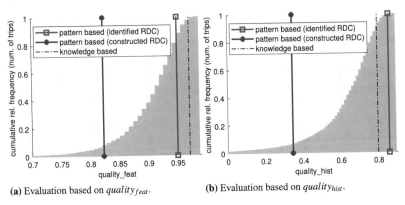

(a) Evaluation based on $quality_{feat}$. **(b)** Evaluation based on $quality_{hist}$.

Figure 5.28 Comparison of pattern- and knowledge-based approaches for RDC construction/identification

5.2.4 Evenly Distributed Pattern Frequency RDCs

In addition to previous analyses of pattern-based RDC construction for use cases 2 or 3, now the possibility of constructing RDCs for use case 1 is evaluated. Hence, the user requires the RDC to be representative for every possible situation in real-life driving, not in its frequency but only in its occurrence. In comparison to use cases 2 and 3 these requirements are easier to accomplish, because the relation of frequency does not need to be maintained.

Three approaches will be compared. First, as proposed in Section 4.3 a cycle will be synthetically constructed by concatenating patterns. In contrast to the previous section, it is much easier to select a representative set of patterns for this construction. Every pattern is supposed to occur exactly once in the RDC. Hence, 255 patterns will be concatenated to construct the RDC. In contrast to the proposal of Section 4.3.2 the patterns are concatenated end to end instead of considering an overlap. Secondly, the trip with the greatest variance of patterns will be selected to be the RDC. This approach is equivalent to the identification of RDCs proposed in Section 4.2. The only difference is, that the selection process is much easier. Because the frequency relations do not need to be considered, the selected trip is the trip that contains every pattern in the best case. Thirdly, the knowledge-based selected trip from Section 5.2.1 is used as a reference, because there is no knowledge-based alternative for this use case.

In Figure 5.29a the results are shown for the occurrence of patterns and the resulting driving-distance of the RDC for all three approaches. For the pattern-based identification, unfortunately, there is no trip that contains every pattern. Hence, the trip that contains the most patterns is chosen to be the RDC. In this case 242 of 255 patterns can be found in the trip. For reference, the knowledge-based RDC contains only 176 patterns. Additionally, the figure shows a logarithmic correlation of distance and contained patterns. Hence, a raise in contained patterns causes a greater jump in distance in high regions of contained patterns than in low regions. This correlation results in a pattern-based identified RDC that has an enormously high distance of 289.7 km and more than 8 hours of driving, which would force the engineers in automotive development to spend a long time with test-bench testing. This amount of time can be reduced by using a semi-synthetically constructed RDC based on patterns. In this case the distance that needed to be driven on test benches can be reduced by approximately one tenth to 34.2 km and 47 minutes of driving time.

In Section 5.2.1 an evaluation criterion for this use case was already introduced. In equivalence to previous analyses Figure 5.29b visualizes the cumulative relative frequency of $quality_{binaryhist}$ for all trips as a bar plot. Additionally, the values for all three RDCs are visualized. The pattern-based identified RDC shows a high quality compared to the knowledge-based RDC. Unfortunately the constructed RDC can not compete with both approaches regarding this evaluation criterion. An impact that cannot be neglected is, that the probability to meet every possible velocity-acceleration state raises when the duration and distance of a cycle is increased. Which is why the results do not surprise. Still, the pattern-based construction approach can be considered as valuable complement to common testing procedures, because the RDC contains all the patterns, which cannot be provided by any other full trip

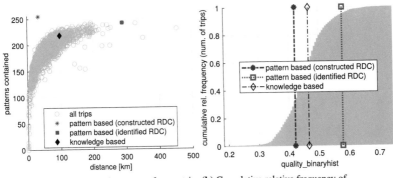

(a) Distance and contained patterns of every trip **(b)** Cumulative relative frequency of
and three RDCs. $quality_{binaryhist}$ for every trip as well as for
                                                                     three RDCs.

Figure 5.29 Three RDCs for use case 1

method. This fact highlights the biggest advantage of constructing an RDC: The
approach adds a degree of freedom to the analysis by giving the user the chance
to individually select a set of patterns that need to be included. In contrast to the
current use case, this can also include patterns that were not discovered by pattern
discovery, but are already known to the user as situations that are very critical to the
component under consideration.

Conclusion

6

Unsupervised pattern discovery deals with the identification of similar recurring subsequences in time series without any knowledge regarding these patterns. The unsupervised discovery of patterns in time series has proven to be beneficial in many different research areas. This thesis investigated the application of pattern discovery in automotive use cases, which was considered as a gap in research previously. The handling of automotive data, or vehicle time series data, is especially challenging because it unites properties like high dynamism, diversity, dimensionality and large sizes. Based on these properties requirements for an appropriate pattern discovery method were derived, which include the following:

- Fast computation of large time series in order to enable the analysis of data sets with more than 100 mio. samples on a regular basis without the need of extraordinary amounts of computing power, as well as the ability to be universally applicable regardless of the kind of data set.
- Discover patterns in multivariate time series occurring simultaneously in each dimension.
- Ability to differentiate between static long-term and dynamic short-term processes, while considering patterns with variable length including time scaling of members of a pattern.
- Invariance regarding amplitudinal scaling and offset, as well as a robustness against noise.
- No necessity of inputting information regarding the number or length of patterns.
- Selection of a set of relevant patterns that in total are representative for the data set.

Based on these requirements, advantages and disadvantages of different approaches for pattern discovery were discussed. The discretization-based approaches were

F. K. D. Noering, *Unsupervised Pattern Discovery in Automotive Time Series*, AutoUni – Schriftenreihe 159, https://doi.org/10.1007/978-3-658-36336-9_6

classified as suitable especially because of their unique ability to process large amounts of data in short time. Unfortunately, this advantage comes with the disadvantage of ignoring too much information in the process of compression. In order to weaken the disadvantages while at the same time maintaining the benefit of fast computation, multiple additional preprocessing functions were proposed (hysteresis, unification, logarithmic reduction). Based on this compressed representation of the time series, two alternative functions were utilized to create a dictionary of patterns. Furthermore, to reduce the amount of discovered patterns and, hence, select a representative set of patterns a parameterizable greedy strategy was proposed. In total, the discretization-based pattern discovery method including all of the proposed supplements is able to match every of the above listed requirements.

All functions were extensively evaluated by means of a procedure to create synthetic time series with labeled patterns and a real-life data set of a battery electric vehicle with a length of approximately 130 mio. samples. The *hysteresis* filtering proved to reduce symbolic toggling and therefore enhanced the quality of discovered patterns (see Section 5.1.1). It even showed better results than common filtering approaches like PAA or Wavelet. In general, filtering turned out to be very important when applying discretization-based pattern discovery. By applying the proposed *logarithmic reduction* technique the pattern discovery is now able to differentiate between a dynamic symbolic value change and a slow long-term value change, which led to an increased precision when rediscovering charging events (see Section 5.1.2). In general, it enables the user to control the amount of time scaling within a pattern. By its logarithmic structure this reduction technique manages the balancing act between enhanced precision and low computing time. This can be achieved by a high compression in cases of long-term static behavior and a low compression in cases of short-term dynamic behavior. In Section 5.1.3 the performance of pattern discovery in an environment with increasing dimensionality was evaluated. Even though the *unification* technique has its limits, the tests showed its ability to discover valid patterns in a five-dimensional time series.

Beyond these additional preprocessing techniques, the greedy postprocessing strategy to select relevant patterns was evaluated in Section 5.1.4 based on a synthetic test case equivalent to the one in Section 5.1.1. The tests showed that the greedy technique is able to reduce the number of patterns to approximately one percent of the initial count, with only a minor decrease in quality. Beside this, two of three weighted greedy techniques (static and dynamic) proved to be very effective in comparison to the standard greedy technique by achieving a higher quality with similar amounts of patterns.

While the whole framework has proven to be very efficient and effective, there is still some potential for further improvement. First, the limitations regarding multivariate pattern discovery should be mentioned here. Due to the compression of multiple dimensions to only one symbolic time series, the approach is not scalable to much larger dimensionality. Future developments will investigate the possibility to apply univariate pattern discovery to each of the dimensions separately and afterwords applying a postprocessing technique to identify relevant multivariate patterns. This approach will open up the opportunity to also discover subdimensional patterns in only a subset of dimensions and the adaption to multidimensional shifting. Beside that, we discovered potential in the selection process of the greedy postprocessing technique. For example in Figure 5.14 the phenomenon of covering samples multiple times was observed, especially when the maximum coverage parameter α is set to high values. This may reduce the ability of the discovered pattern set to be representative for the data set. The selection process could be enhanced to prevent or weaken this phenomenon. Last but not least, a potential improvement is to reduce the effort of parametrization. Especially the number and location of the discretization boundaries has a high influence on the results of the pattern discovery regarding the level of detail of patterns. In many cases these automatically set boundaries can be enhanced by the expert. In order to increase the level of independence of expert knowledge, it may be beneficial to enhance the setting of discretization boundaries.

In order to show the potential benefit of using the pattern discovery it was applied to the construction of representative cycles, which are commonly used to represent the real-life usage of technical devices in a compressed way. The biggest challenge in creating representative cycles is to define the representativeness. In Section 2.2.2 various possibilities to define the representativeness for different use cases were reviewed. What they all have in common is that they are based on features contributed by experts, so called characteristic parameters (CP). These CPs are customized for each use case. Hence, the user has to put a lot of effort into defining these CPs. Beside this, another drawback is that these CPs can only capture a limited part of the real life, because they are highly simplified. Based on the CPs a cycle is chosen or constructed that best matches the CP values of the original data.

In this thesis a new pattern-based approach of constructing representative cycles was introduced that eliminates the need of defining customized CPs. Therefore, pattern-based statistics like the *Absolute Pattern Frequency* (APF) and the *Relative Pattern Coverage* (RPC) are introduced. Based on the APF- or RPC-spectrum it is now possible to characterize a data set. Furthermore, distance metrics were introduced to measure the similarity between data sets based on their APF- or RPC-spectra. Hence, the CPs are not defined by experts but implicitly by the pattern discovery. This approach is enabled by the ability of the pattern discovery and its greedy postprocessing strategy to select a representative pattern set. Based on these pattern-based statistics two alternative approaches to construct representative cycles were introduced. First, a full trip method was proposed, which selects the cycle that best matches the APF- or RPC-spectrum of the input data set. Secondly, an approach was introduced to construct a representative cycle by concatenating patterns under consideration of a perfect APF-spectrum and a representative pattern set. In order to evaluate the newly introduced approaches, both methods were applied to derive representative driving cycles (RDC) based on a fleet of Volkswagen vehicles that in total drove approximately 172,000 km in 2700 trips. Furthermore, three evaluation criteria were introduced to assess the performance of RDCs, while the first one was also used to derive a knowledge-based RDC as a reference for the newly proposed techniques.

In terms of the first pattern-based RDC identification technique, two challenges could be identified. First of all, the curse of dimensionality when comparing two data sets in the APF or RPC space complicates the task of choosing the nearest neighbor and, hence, identifying the RDC. This is why the possibility to use fractional distance metrics instead of Euclidean or Manhattan distance was investigated. As described by the authors of [2], the meaningfulness of fractional distance metrics ought to worsen slower when increasing the dimensionality in comparison to Euclidean or Manhattan distance metrics. In contrast to the assessment of [2], the Euclidean distance proved to be the most stable distance metric when increasing the dimensionality, which goes along with findings of Klawonn et al. in [48]. Secondly, when increasing the maximum coverage parameter α the proportion of patterns having low values of APF or RPC is increasing. This is especially problematic when calculating the distances. In order to solve this issue, a normalization technique was proposed. The experiments showed that the normalization technique has a positive impact on the selection process. In general, it was proven that the pattern-based RDC identification approach can compete with the knowledge-based reference. Furthermore, regarding the independent evaluation criterion 2 the pattern-based approach even exceeded the quality of the knowledge-based reference. For validation the pattern discovery parameters were slightly adapted, which resulted in even better RDC quality values.

In contrast to the pattern-based full trip approach, the proposed pattern-based RDC construction technique is much more complex. Based on an optimal APF-spectrum a pattern set for the RDC is composed, whose patterns are concatenated later. In order to identify the best sequence of patterns, the problem is translated into an Asymmetric Traveling Salesman problem (ATSP). This problem is classified as NP-hard and is solved by a heuristic based on Linear Programming (LP). In general, this approach has two issues. First, the complexity of the ATSP, which is solved by the LP heuristic. Secondly, the challenge of creating cycles that are physically valid and that could be driven on e.g. test benches. For the later issue multiple techniques to transform the selected sequence of patterns into a physically valid driving cycle are proposed in this thesis. The physical validity was proven by means of evaluation criterion 3.

Even though the complexity of ATSP limits the number of patterns to be included in an RDC, this approach has shown the potential of creating the shortest possible RDC that includes all the real-life situations. As proven previously, this property can not be provided by any other full trip method. On the other hand, the proposed pattern-based full trip approach was able to identify RDCs that are at least as representative as the knowledge-based reference, without actually inputting information regarding the knowledge-based objectives.

To conclude this, pattern-based approaches can be considered in various use cases as valuable supplements or even alternatives. Hence, with its ability to process large amounts of data, the unsupervised pattern discovery is able to add a data-driven view to the engineering perspective, which boosts the knowledge of the real-life usage and subsequently opens up the possibility for product-improvement.

References

1. Reducing car emissions: new CO2 targets for cars explained. https://www.europarl.europa.eu/news/en/headlines/society/20180920STO14027/reducing-car-emissions-new-co2-targets-for-cars-explained (Apr 2019), accessed: 2021-02-04
2. Aggarwal, C.C., Hinneburg, A., Keim, D.A.: On the Surprising Behavior of Distance Metric in High-Dimensional Space. In: Van den Bussche, J., Vianu, V. (eds.) Database Theory—ICDT 2001. pp. 420–434. No. 1973 in Lecture Notes in Computer Science, Springer Berlin Heidelberg, Berlin, Heidelberg (Feb 2001)
3. Akhter, F.: A Heuristic Approach for Minimum Set Cover Problem. International Journal of Advanced Research in Artificial Intelligence 4 (Jun 2015)
4. Allen, J.F.: Maintaining Knowledge about Temporal Intervals. Commun. ACM 26(11), 832–843 (Nov 1983)
5. Andre, M.: The ARTEMIS European driving cycles for measuring car pollutant emissions. The Science of the total environment 334–335, 73–84 (Jan 2005)
6. Azulay, R., Moskovitch, R., Stopel, D., Verduijn, M., Jonge, E., Shahar, Y.: Temporal Discretization of medical time series—A comparative study (Jan 2007)
7. Balasubramanian, A., Prabhakaran, B.: Flexible Exploration and Visualization of Motifs in Biomedical Sensor Data. In: Workshop on Data Mining for Healthcare (DMH), ACM Conference of Knowledge Discovery and Data Mining (Aug 2013)
8. Balasubramanian, A., Wang, J., Prabhakaran, B.: Discovering Multidimensional Motifs in Physiological Signals for Personalized Healthcare. IEEE Journal of Selected Topics in Signal Processing 10 (Aug 2016)
9. Bashar, M.A., Li, Y.: Interpretation of text patterns. Data Mining and Knowledge Discovery (Feb 2018)
10. Castro, N., Azevedo, P.: Multiresolution Motif Discovery in Time Series. pp. 665–676 (Apr 2010)
11. Chandu, D.P.: Improved Greedy Algorithm for Set Covering Problem (2015)
12. Chiu, B., Keogh, E., Lonardi, S.: Probabilistic discovery of time series motifs. In: Proceedings of the ninth ACM SIGKDD international conference on Knowledge discovery and data mining. pp. 493–498. ACM (2003)
13. Chvatal, V.: A Greedy Heuristic for the Set-Covering Problem. Mathematics of Operations Research 4(3), 233–235 (Aug 1979)
14. Dai, Z., Niemeier, D., Eisinger, D.: Driving cycles: a new cycle-building method that better represents real-world emissions (Jan 2008)

15. Dakin, R.J.: A tree-search algorithm for mixed integer programming problems. The Computer Journal 8(3), 250–255 (Jan 1965)
16. Dau, H.A., Keogh, E.: Matrix Profile V: A Generic Technique to Incorporate Domain Knowledge into Motif Discovery. In: Proceedings of the 23rd ACM SIGKDD International Conference on Knowledge Discovery and Data Mining. pp. 125–134. KDD '17, Association for Computing Machinery, New York, NY, USA (2017)
17. Deppe, S., Lohweg, V.: Shift-Invariant Feature Extraction for Time-Series Motif Discovery (Nov 2015)
18. Esmael, B., Arnaout, A., Fruhwirth, R.K., Thonhauser, G.: Multivariate Time Series Classification by Combining Trend-Based and Value-Based Approximations. In: Computational Science and Its Applications—ICCSA 2012. pp. 392–403 (2012)
19. Esser, A., Kohnhäuser, F., Ostern, N., Engleson, K., Rinderknecht, S.: Enabling a Privacy-Preserving Synthesis of Representative Driving Cycles from Fleet Data using Data Aggregation. In: 2018 21st International Conference on Intelligent Transportation Systems (ITSC). pp. 1384–1389 (Nov 2018)
20. Eßer, A., Zeller, M., Foulard, S., Rinderknecht, S.: Stochastic Synthesis of Representative and Multidimensional Driving Cycles. SAE International Journal of Alternative Powertrains 7(3), 263–272 (Apr 2018)
21. Fries, M., Baum, A., Wittmann, M., Lienkampk, M.: Derivation of a real-life driving cycle from fleet testing data with the Markov-Chain-Monte-Carlo Method. In: 2018 21st International Conference on Intelligent Transportation Systems (ITSC). pp. 2550–2555 (Nov 2018)
22. Fujito, T., Okumura, T.: A Modified Greedy Algorithm for the Set Cover Problem with Weights 1 and 2. pp. 670–681 (Dez 2001)
23. Galgamuwa, U., Perera, L., Bandara, S.: Developing a General Methodology for Driving Cycle Construction: Comparison of Various Established Driving Cycles in the World to Propose a General Approach. Journal of Transportation Technologies 05, 191–203 (Jan 2015)
24. Gao, Y., Lin, J.: Exploring variable-length time series motifs in one hundred million length scale. Data Mining and Knowledge Discovery 32(5), 1200–1228 (2018)
25. Gao, Y., Lin, J., Rangwala, H.: Iterative Grammar-Based Framework for Discovering Variable-Length Time Series Motifs. In: 15th IEEE International Conference on Machine Learning and Applications (ICMLA). pp. 7–12. IEEE (2016)
26. Gao, Y., Lin, J.: HIME: discovering variable-length motifs in large-scale time series. Knowledge and Information Systems 61 (Dez 2018)
27. Gharghabi, S., Ding, Y., Yeh, C.M., Kamgar, K., Ulanova, L., Keogh, E.: Matrix Profile VIII: Domain Agnostic Online Semantic Segmentation at Superhuman Performance Levels. pp. 117–126 (Nov 2017)
28. Gharghabi, S., Imani, S., Bagnall, A.J., Darvishzadeh, A., Keogh, E.J.: Matrix Profile XII: MPdist: A Novel Time Series Distance Measure to Allow Data Mining in More Challenging Scenarios. In: IEEE International Conference on Data Mining, ICDM 2018, Singapore, November 17–20, 2018. pp. 965–970 (2018)
29. Gong, H., Zou, Y., Yang, Q., Fan, J., Sun, F., Goehlich, D.: Generation of a driving cycle for battery electric vehicles: A case study of Beijing. Energy 150, 901–912 (2018)
30. Grabocka, J., Schilling, N., Schmidt-Thieme, L.: Latent Time-Series Motifs. ACM Transactions on Knowledge Discovery from Data 11, 1–20 (Jul 2016)

31. Grossman, T., Wool, A.: Computational experience with approximation algorithms for the set covering problem. European Journal of Operational Research 101(1), 81–92 (1997)

32. Gwo-Hshiung, T., June-Jye, C.: Developing A Taipei motorcycle driving cycle for emissions and fuel economy. Transportation Research Part D: Transport and Environment 3(1), 19–27 (1998)

33. Günnel, T.: NEFZ, WLTP, RDE und PEMS—ein Überblick (2019), https://www.automobil-industrie.vogel.de/nefz-wltp-rde-und-pems-ein-ueberblick-a-657992/, accessed: 2021-02-04

34. Hallac, D., Vare, S., Boyd, S., Leskovec, J.: Toeplitz Inverse Covariance-Based Clustering of Multivariate Time Series Data. In: Proceedings of the 23rd ACM SIGKDD International Conference on Knowledge Discovery and Data Mining. pp. 215–223. KDD '17, Association for Computing Machinery, New York, NY, USA (2017)

35. Heinrich, F., Noering, F.K.D., Pruckner, M., Jonas, K.: Unsupervised Datapreprocessing for LSTM-based Battery Model under Electric Vehicle Operation. Journal of Energy Storage (2021), currently under review

36. Ho, S., Wong, Y., Chang, V.W.C.: Developing Singapore Driving Cycle for passenger cars to estimate fuel consumption and vehicular emissions. Atmospheric Environment 97, 353–362 (2014)

37. Hoeppner, F.: Knowledge Discovery from Sequential Data. Ph.D. thesis, Technical University Braunschweig (2003)

38. Hongwen, H., Jinquan, G., Jiankun, P., Huachun, T., Chao, S.: Real-time global driving cycle construction and the application to economy driving pro system in plug-in hybrid electric vehicles. Energy 152 (Mar 2018)

39. Huertas, J., Giraldo, M., Quirama, L., Díaz-Ramírez, J.: Driving Cycles Based on Fuel Consumption. Energies 11, 3064 ff. (Nov 2018)

40. Huertas, J., Quirama, L., Giraldo, M., Díaz-Ramírez, J.: Comparison of driving cycles obtained by the Micro-trips, Markov- chains and MWD-CP methods (Aug 2019)

41. Huertas, J., Quirama, L., Giraldo, M., Díaz-Ramírez, J.: Comparison of Three Methods for Constructing Real Driving Cycles. Energies 12, 665 ff. (Feb 2019)

42. Imani, S., Keogh, E.: Matrix Profile XVI: Time Series Semantic Motifs: A New Primitive for Finding Higher-Level Structure in Time Serie. In: IEEE ICDM 2019 (2019)

43. Imani, S., Madrid, F., Ding, W., Crouter, S.E., Keogh, E.J.: Matrix Profile XIII: Time Series Snippets: A New Primitive for Time Series Data Mining. In: 2018 IEEE International Conference on Big Knowledge, ICBK 2018, Singapore, November 17–18, 2018. pp. 382–389 (2018)

44. Jancovic, P., Köküer, M., Zakeri, M., Russel, M.: Unsupervised discovery of acoustic patterns in bird Vocalisations employing DTW and clustering. In: 21st European Signal Processing Conference. pp. 285–296 (2013)

45. Kamgar, K., Gharghabi, S., Keogh, E.J.: Matrix Profile XV: Exploiting Time Series Consensus Motifs to Find Structure in Time Series Sets. In: IEEE ICDM 2019 (2019)

46. Kane, A., Shiri, N.: Multivariate Time Series Representation and Similarity Search Using PCA. In: Advances in Data Mining—Applications and Theoretical Aspects. pp. 122–136 (2017)

47. Karp, R.M.: Reducibility Among Combinatorial Problems. Complexity of Computer Computations pp. 85–103 (1972)

48. Klawonn, F., Höppner, F., Jayaram, B.: What are Clusters in High Dimensions and are they Difficult to Find? In: Masulli, F., Petrosino, A., Rovetta, S. (eds.) Clustering High-Dimensional Data—First International Workshop, CHDD 2012, Naples, Italy, May 15, 2012, Revised Selected Papers. Lecture Notes in Computer Science, vol. 7627, pp. 14–33. Springer (2012)

49. Lee, T., Adornato, B., Filipi, Z.S.: Synthesis of Real-World Driving Cycles and Their Use for Estimating PHEV Energy Consumption and Charging Opportunities: Case Study for Midwest/U.S. IEEE Transactions on Vehicular Technology 60(9), 4153–4163 (Nov 2011)

50. Lee, T., Filipi, Z.S.: Synthesis and validation of representative real-world driving cycles for Plug-In Hybrid vehicles. In: 2010 IEEE Vehicle Power and Propulsion Conference. pp. 1–6 (Sep 2010)

51. Lee, T.K., Filipi, Z.: Synthesis of real-world driving cycles using stochastic process and statistical methodology. Int. J. of Vehicle Design 57, 17–36 (Nov 2011)

52. Li, Y., Lin, J., Oates, T.: Visualizing Variable-Length Time Series Motifs. In: Proceedings of the Twelfth SIAM International Conference on Data Mining, Anaheim, California, USA, April 26–28, 2012. pp. 895–906 (2012)

53. Liessner, R., Fechert, R., Bäker, B.: Derivation of Real Driving Emission Cycles based on Real-world Driving Data—Using Markov Models and Threshold Accepting. pp. 188–195 (Jan 2017)

54. Lin, J., Keogh, E., Lonardi, S., Chiu, B.: A symbolic representation of time series, with implications for streaming algorithms. In: ACM SIGMOD Workshop on Research Issues in Data Mining and Knowledge Discovery (DMKD'03) (2003)

55. Lin, J., Niemeier, D.: An exploratory analysis comparing a stochastic driving cycle to California's regulatory cycle. Atmospheric Environment 36, 5759–5770 (Dez 2002)

56. Lin, J., Niemeier, D.: Estimating Regional Air Quality Vehicle Emission Inventories: Constructing Robust Driving Cycles. Transportation Science 37, 330–346 (Aug 2003)

57. Lin, J., Li, Y.: Finding approximate frequent patterns in streaming medical data. pp. 13–18 (Nov 2010)

58. Linardi, M., Zhu, Y., Palpanas, T., Keogh, E.J.: Matrix Profile X: VALMOD—Scalable Discovery of Variable-Length Motifs in Data Series. In: Proceedings of the 2018 International Conference on Management of Data, SIGMOD Conference 2018, Houston, TX, USA, June 10–15, 2018. pp. 1053–1066 (2018)

59. von der Lippe, P., Kladroba, A.: Repräsentativität von Stichproben. Marketing ZFP 24, 139–145 (2002)

60. Madrid, F., Imani, S., Mercer, R., Zimmerman, Z., Shakibay, N., Keogh, E.: Matrix Profile XX: Finding and Visualizing Time Series Motifs of All Lengths using the Matrix Profile. In: IEEE Big Knowledge 2019 (2019)

61. Madrid, F., Singh, S., Chesnais, Q., Mauck, K., Keogh, E.: Matrix Profile XIX: Efficient and Effective Labeling of Massive Time Series Archives. In: DSAA 2019: International Conference on Data Science and Advanced Analytics (2019)

62. Mahayadin, A.R., Ibrahim, I., Zunaidi, I., Shahriman, A.B., Faizi, M.K., Sahari, M., Hashim, M.S.M., Saad, M.A.M., Sarip, M.S., Razlan, Z.M., Rani, M.F.H., Isa, Z.M., Kamarrudin, N.S., Harun, A., Nagaya, Y.: Development of Driving Cycle Construction Methodology in Malaysia's Urban Road System. In: 2018 International Conference

on Computational Approach in Smart Systems Design and Applications (ICASSDA). pp. 1–5 (Aug 2018)

63. Minnen, D., Isbell, C., Essa, I., Starner, T.: Detecting Subdimensional Motifs: An Efficient Algorithm for Generalized Multivariate Pattern Discovery. In: 7th IEEE International Conference on Data Mining. pp. 601–606 (2007)

64. Minnen, D., Isbell, C.L., Essa, I., Starner, T.: Discovering Multivariate Motifs Using Subsequence Density Estimation and Greedy Mixture Learning. In: Proceedings of the 22nd National Conference on Artificial Intelligence—Volume 1. pp. 615–620. AAAI'07, AAAI Press (2007)

65. Mörchen, F., Ultsch, A., Hoos, O.: Extracting interpretable muscle activation patterns with time series knowledge mining. KES Journal 9, 197–208 (2005)

66. Mueen, A.: Enumeration of Time Series Motifs of All Lengths. In: 2013 IEEE 13th International Conference on Data Mining, Dallas, TX, USA, December 7–10, 2013. pp. 547–556 (2013)

67. Mueen, A., Keogh, E., Zhu, Q., Cash, S., Westover, M.B.: Exact Discovery of Time Series Motifs. vol. 2009, pp. 473–484 (Apr 2009)

68. Nelder, J.A., Mead, R.: A Simplex Method for Function Minimization. Comput. J. 7, 308–313 (1965)

69. Nevill-Manning, C., Witten, I.: Identifying hierarchical structure in sequences: a linear time algorithm. Journal of Artificial Intelligence Research 7, 67–82 (1997)

70. Noering, F.K.D., Jonas, K., Klawonn, F.: Assessment and Adaption of Pattern Discovery Approaches for Time Series Under the Requirement of Time Warping. In: Proceedings of 19th Intelligent Data Engineering and Automated Learning (IDEAL'18). LNCS, vol. 11314, pp. 285–296. Springer International Publishing (2018)

71. Noering, F.K.D., Jonas, K., Klawonn, F.: Improving Discretization Based Pattern Discovery for Multivariate Time Series by Additional Preprocessing. Intelligent Data Analysis 25(5) (Sep 2021), accepted for publication

72. Noering, F.K.D., Schroeder, Y., Jonas, K., Klawonn, F.: Pattern Discovery in Time Series Using Autoencoder in Comparison to Nonlearning Approaches. Integrated Computer-Aided Engineering (2021), accepted for publication

73. Nunthanid, P., Niennattrakul, V., Ratanamahatana, C.: Discovery of variable length time series motif. pp. 472–475 (May 2011)

74. Nunthanid, P., Niennattrakul, V., Ratanamahatana, C.A.: Parameter-free motif discovery for time series data. In: 2012 9th International Conference on Electrical Engineering/Electronics, Computer, Telecommunications and Information Technology. pp. 1–4 (May 2012)

75. Nyberg, P., Frisk, E., Nielsen, L.: Using Real-World Driving Databases to Generate Driving Cycles With Equivalence Properties. IEEE Transactions on Vehicular Technology 65(6), 4095–4105 (Jun 2016)

76. Rakthanmanon, T., Campana, B., Mueen, A., Batista, G., Westover, M.B., Zhu, Q., Zakaria, J., Keogh, E.: Searching and Mining Trillions of Time Series Subsequences under Dynamic Time Warping. vol. 2012 (Aug 2012)

77. Rayan, Y., Mohammad, Y., Ali, S.: Multidimensional Permutation Entropy for Constrained Motif Discovery, pp. 231–243 (Jan 2019)

78. Ritt, R., O'Leary, P.: Symbolic Analysis of Machine Behaviour and the Emergence of the Machine Language. In: Theory and Practice of Natural Computing—7th Interna-

tional Conference,TPNC 2018, Dublin, Ireland, December 12–14, 2018, Proceedings. pp. 305–316 (2018)

79. Rupasinghe, H., Rengarasu, T.: Development of Driving Cycles for Galle. pp. 108–113 (May 2018)

80. Seers, P., Nachin, G., Glaus, M.: Development of two driving cycles for utility vehicles. Transportation Research Part D: Transport and Environment 41(C), 377–385 (2015)

81. Senin, P., Lin, J., Wang, X., Oates, T., Gandhi, S., Boedihardjo, A., Chen, C., Frankenstein, S., Lerner, M.: GrammarViz 2.0: A Tool for Grammar-Based Pattern Discovery in Time Series. Machine Learning and Knowledge Discovery in Databases 8726, 468–472 (2014)

82. Senin, P., Lin, J., Wang, X., Oates, T., Gandhi, S., Boedihardjo, A.P., Chen, C., Frankenstein, S.: GrammarViz 3.0: Interactive Discovery of Variable-Length Time Series Patterns. ACM Trans. Knowl. Discov. Data 12(1) (Feb 2018)

83. Shahidinejad, S., Bibeau, E., Filizadeh, S.: Statistical Development of a Duty Cycle for Plug-in Vehicles in a North American Urban Setting Using Fleet Information. IEEE Transactions on Vehicular Technology 59(8), 3710–3719 (Oct 2010)

84. Shakibay Senobari, N., Funning, G., Zimmerman, Z., Zhu, Y., Keogh, E.: Using the similarity Matrix Profile to investigate foreshock behavior of the 2004 Parkfield earthquake. In: American Geophysical Union, Fall Meeting 2018 (2018)

85. Shieh, J., Keogh, E.: iSAX: disc-aware mining and indexing of massive time series data. In: Data Mining and Knowledge Discovery (2009)

86. Shieh, J., Keogh, E.: ISAX: Indexing and mining terabyte sized time series. Proceedings of the ACM SIGKDD International Conference on Knowledge Discovery and Data Mining pp. 623–631 (Aug 2008)

87. Silva, D.F., Yeh, C.C.M., Batista, G.E.A.P.A., Keogh, E.J.: SiMPle: Assessing Music Similarity Using Subsequences Joins. In: ISMIR (2016)

88. Son, N.T., Anh, D.T.: Discovery of time series k-motifs based on multidimensional index. Knowledge and Information Systems 46 (Jan 2015)

89. Sun, C., Stirling, D., Ritz, C., Sammut, C.: Variance-wise segmentation for a temporal-adaptive sax. In: Zhao, Y., Li, J., Kennedy, P., Christen, P. (eds.) Data Mining and Analytics 2012 (AusDM 2012). CRPIT, vol. 134, pp. 71–78. ACS, Sydney, Australia (2012)

90. Tanaka, Y., Iwamoto, K., Uehara, K.: Discovery of Time-Series Motif from Multi-Dimensional Data Based on MDL Principle. In: Machine Learning. pp. 269–300 (2005)

91. Tanaka, Y., Uehara, K.: Discover Motifs in Multi-dimensional Time-Series Using the Principal Component Analysis and the MDL Principle. vol. 2734, pp. 252–265 (Jul 2003)

92. Tharvin, R., Kamarrudin, N.S., bakar, S.A., Ibrahim, Z., Razlan, Z.M., Khairunizam, W., Harun, A., Hashim, M., Ibrahim, I., Rahman, M.K.F.A., Saad, M., Mahayadin, A., Hilmi, M.F.: Development of Driving Cycle for Passenger Car under Real World Driving Conditions in Kuala Lumpur, Malaysia. IOP Conference Series: Materials Science and Engineering 429 (Nov 2018)

93. Tiakas, E., Papadopoulos, A.N., Manolopoulos, Y.: Skyline queries: An introduction. In: 6th International Conference on Information, Intelligence, Systems and Applications, IISA 2015, Corfu, Greece, July 6–8, 2015. pp. 1–6. IEEE (2015)

94. Toyoda, M., Sakurai, Y., Ishikawa, Y.: Pattern discovery in data streams under the time warping distance. Very Large Data Bases 22(3), 295–318 (2013)

95. Vahdatpour, A., Amini, N., Sarrafzadeh, M.: Toward Unsupervised Activity Discovery Using Multi-Dimensional Motif Detection in Time Series. In: 21st International Joint Conference on Artificial Intelligence. vol. 9, pp. 1261–1266 (2009)

96. Wadephul, C.: Sind Heuristiken die besseren Algorithmen? Ein Antwortversuch am Beispiel des Traveling Salesman Problem (TSP), pp. 55–93. Springer Verlag (2020)

97. Wang, Z., Zhang, J., Liu, P., Qu, C., Li, X.: Driving Cycle Construction for Electric Vehicles Based on Markov Chain and Monte Carlo Method: A Case Study in Beijing. Energy Procedia 158, 2494–2499 (2019), innovative Solutions for Energy Transitions

98. Xu, H., Ou, Z.: Scalable Discovery of Audio Fingerprint Motifs in Broadcast Streams With Determinantal Point Process Based Motif Clustering. IEEE/ACM Transactions on Audio, Speech, and Language Processing 24(5), 978–989 (2016)

99. Yankov, D., Keogh, E., Medina, J., Chiu, B., Zordan, V.: Detecting Time Series Motifs under Uniform Scaling. In: Proceedings of the 13th ACM SIGKDD International Conference on Knowledge Discovery and Data Mining. pp. 844–853. KDD '07, Association for Computing Machinery, New York, NY, USA (2007)

100. Ye, L., Keogh, E.: Time series shapelets: a new primitive for data mining. pp. 947–956 (Jul 2009)

101. Yeh, C.M., Herle, H.V., Keogh, E.J.: Matrix Profile III The Matrix Profile Allows Visualization of Salient Subsequences in Massive Time Series. In: IEEE 16th International Conference on Data Mining, ICDM 2016, December 12–15, 2016, Barcelona, Spain. pp. 579–588 (2016)

102. Yeh, C.M., Kavantzas, N., Keogh, E.: Matrix Profile IV: Using Weakly Labeled Time Series to Predict Outcomes. Proc. VLDB Endow. 10(12), 1802–1812 (Aug 2017)

103. Yeh, C.M., Kavantzas, N., Keogh, E.: Matrix Profile VI: Meaningful Multidimensional Motif Discovery. In: IEEE International Conference. pp. 565–574 (2017)

104. Yeh, C.M., Zhu, Y., Ulanova, L., Begum, N., Ding, Y., Dau, A., Silva, D., Mueen, A., Keogh, E.: Matrix Profile I: All Pairs Similarity Joins for Time Series: A Unifying View That Includes Motifs, Discords and Shapelets. pp. 1317–1322 (Dez 2016)

105. Yingchareonthawornchai, S., Sivaraks, H., Rakthanmanon, T., Ratanamahatana, C.: Efficient Proper Length Time Series Motif Discovery. pp. 1265–1270 (Dez 2013)

106. Yuhui, P., Zhuang, Y., Yang, Y.: A driving cycle construction methodology combining k-means clustering and Markov model for urban mixed roads. Proceedings of the Institution of Mechanical Engineers, Part D: Journal of Automobile Engineering (May 2019)

107. Zaccardi, J.M., Berr, F.L.: Analysis and choice of representative drive cycles for light duty vehicles—case study for electric vehicles (2013)

108. Zhang, M., Shi, S., Lin, N., Yue, B.: High-Efficiency Driving Cycle Generation Using a Markov Chain Evolution Algorithm. IEEE Transactions on Vehicular Technology 68(2), 1288–1301 (Feb 2019)

109. Zhao, J., Gao, Y., Guo, J., Chu, L.: The creation of a representative driving cycle based on Intelligent Transportation System (ITS) and a mathematically statistical algorithm: A case study of Changchun (China). Sustainable Cities and Society 42, 301–313 (2018)

110. Zhao, X., Ma, J., Wang, S., Ye, Y., Wu, Y., Yu, M.: Developing an electric vehicle urban driving cycle to study differences in energy consumption. Environmental Science and Pollution Research 26 (Nov 2018)

111. Zhao, X., Yu, Q., Ma, J., Wu, Y., Yu, M., Ye, Y.: Development of a Representative EV Urban Driving Cycle Based on a k-Means and SVM Hybrid Clustering Algorithm. Journal of Advanced Transportation 2018, 1–18 (Nov 2018)

112. Zhu, Y., Imamura, M., Nikovski, D., Keogh, E.J.: Matrix Profile VII: Time Series Chains: A New Primitive for Time Series Data Mining. In: 2017 IEEE International Conference on Data Mining, ICDM 2017, New Orleans, LA, USA, November 18–21, 2017. pp. 695–704 (2017)

113. Zhu, Y., Yeh, C.M., Zimmerman, Z., Kamgar, K., Keogh, E.J.: Matrix Profile XI: SCRIMP++: Time Series Motif Discovery at Interactive Speeds. In: IEEE International Conference on Data Mining, ICDM 2018, Singapore, November 17–20, 2018. pp. 837–846 (2018)

114. Zhu, Y., Zimmerman, Z., Senobari, N.S., Yeh, C.M., Funning, G., Mueen, A., Brisk, P., Keogh, E.J.: Matrix Profile II: Exploiting a Novel Algorithm and GPUs to Break the One Hundred Million Barrier for Time Series Motifs and Joins. In: IEEE International Conference on DataMining. pp. 739–748 (2016)

115. Zhu, Y., Gharghabi, S., Silva, F.D., Dau, A.H., Yeh, M.C.C., Senobari, S.N., Almaslukh, A., Kamgar, K., Zimmerman, Z., Funning, G., Mueen, A., Keogh, E.: The Swiss Army Knife of Time Series Data Mining: Ten Useful Things you can do with the Matrix Profile and Ten Lines of Code. Data Mining and Knowledge Discovery (2020)

116. Zimmerman, Z., Kamgar, K., Senobari, N.S., Crites, B., Funning, G., Brisk, P., Keogh, E.J.: Matrix Profile XIV: Scaling Time Series Motif Discovery with GPUs to Break a Quintillion Pairwise Comparisons a Day and Beyond. In: Proceedings of the ACM Symposium on Cloud Computing, SoCC 2019, Santa Cruz, CA, USA, November 20–23, 2019. pp. 74–86 (2019)

Printed in the United States
by Baker & Taylor Publisher Services